What people

Universe Within

Our understanding of the Universe, and our place in it, advances in fits and starts when scientists and engineers place seemingly unrelated ideas together and ask, with an open mind, what new things can be learned from the juxtaposition. This book does just that in comparing what is known about the large-scale structure of the Universe to what is known about the structure of the human brain. The author thoroughly explains this comparison without resorting to excessive magic or excessively complicated mathematics. The result is an imaginative and enjoyable exposé covering a broad range of science. Books such as this one are often just the ticket to encourage new thought and scholarship along one of the garden paths that advance our understanding of our place in the Universe. This is a good read, one that I will assign to my physical science students for a semester-long reading and writing exercise.

Dr. Calvin W. Lowe, Professor of Physics and Director of the Interdisciplinary STEM Center at Hampton University, Hampton, VA (USA)

Mr. Felton takes us on a heady, if sometimes dizzying, tour through the wondrous and fantastical landscapes of humankind's outer and inner worlds. Drawing richly on perspectives in physics and neuroscience that range from the well-established to the currently emerging, *Universe Within* covers enormous empirical, conceptual, and theoretical grounds; from fundamental forces and particles to neural networks in the brain and, ultimately, to profound questions of our religion and philosophy. The author takes an intellectual stand that is both bold and brave, posing an intriguing hypothesis about the potentially deep connections between the structure and dynamics of the physical universe and the physical brains that inhabit it in our little corner of reality. Well-written and engaging throughout, Mr. Felton's book is at once both thoroughly accessible and genuinely controversial, containing ideas on nearly every page that will stretch the reader's knowledge and thinking about the world, both as it is and as it might yet turn out to be. From cover to cover, it is an enjoyable and thought-provoking read.

Dr. Kelvin S. Oie, Cognitive Neuroscientist

Melvin's essential point—namely, the structural and dynamical similarities between the brain and the universe at large, which he accounts for by arguing that the brain physically models the universe in its own image—is one that deserves very serious consideration from the academic community and readers interested in a deeper, truer understanding of the nature of what we call reality. This book is a necessary step in the direction of exploring this crucial realization, whose implications are simultaneously vast and profound.

Bernardo Kastrup

Universe Within

The Surprising Way the Human Brain
Models the Universe

Universe Within

The Surprising Way the Human Brain Models the Universe

Melvin A. Felton, Jr.

IFF
BOOKS

Winchester, UK
Washington, USA

JOHN HUNT PUBLISHING

First published by iff Books, 2021
iff Books is an imprint of John Hunt Publishing Ltd., No. 3 East Street, Alresford,
Hampshire SO24 9EE, UK
office@jhpbooks.com
www.johnhuntpublishing.com
www.iff-books.com

For distributor details and how to order please visit the 'Ordering' section on our website.

ISBN: 978 1 78904 719 6
978 1 78904 720 2 (ebook)
Library of Congress Control Number: 2020947594

A CIP catalogue record for this book is available from the British Library.

Design: Stuart Davies

UK: Printed and bound by CPI Group (UK) Ltd, Croydon, CR0 4YY
Printed in North America by CPI GPS partners

We operate a distinctive and ethical publishing philosophy in
all areas of our business, from our global network of authors to
production and worldwide distribution.

Contents

List of Figures

Figure 1: The systems of the universe categorized based on their degree of spatiotemporal extent and level of complexity. This figure encodes complexity in the font color of the level's title—lighter shades for less complex systems and darker shades for more complex ones.

Figure 2: Stages of neutron to proton beta decay. "N" – neutron, "P" – proton, "u" – up-quark, "d" – down-quark, "v_e" – electron neutrino, "e" – electron. The arrow connected to a particle indicates that the particle has kinetic energy and is therefore traveling through space.

Figure 3: The electron double-slit experiment. (a) Electrons depicted in particle form propagating through the experimental apparatus and impacting the detector screen in locations that correspond to peaks in an interference pattern. (b) An electron probability wave propagating through the experimental apparatus and producing an interference pattern at the detector screen. (c) From top to bottom, the buildup of the interference pattern as the intensity of the electron beam is increased. (By user: Belsazar – Provided with kind permission of Dr. Tonomura, CC BY-SA 3.0, https://commons.wikimedia.org/w/index.php?curid=498735)

Figure 4: Simulated large-scale distribution of matter and dark matter throughout the universe—the cosmic web of matter. The brighter regions are extremely large clusters of galaxies, long linear arrays of galaxies called filaments, and 2D galactic systems called walls. The darker regions are expanding voids. (CC BY-SA 4.0, https://commons.wikimedia.org/w/index.php?curid=84870202)

the neocortical hierarchy. Neocortical layer is indicated by the Roman numerals on the left. For simplicity, only a small subset of the approximately 142 neurons that are in a minicolumn are shown—just a few pyramidal and stellate neurons. Also for simplicity, all axons and most of the detail in the dendrites are omitted.

Figure 11: Neocortical hierarchy and its input from the thalamus and hippocampus.

Figure 12: Minicolumns join to form larger systems generally referred to as assemblies, depicted here as consecutive darkened minicolumns. Multiple assemblies, some on different levels of the neocortical hierarchy, can form even larger systems of neurons called coalitions. The assemblies of these coalitions tend to be linked via lateral, feedforward, and feedback long-range projections.

Figure 13: Thalamocortical input. (a) Laminar distribution of the two types of thalamocortical projections—nonspecific and specific. (b) The thalamus often plays a critical role in the formation of neocortical coalitions.

Figure 14: The hippocampus communicates extensively with the intermediate and frontal levels of the neocortical hierarchy.

Figure 15: Large-scale anatomy and information flow in neocortical visual processing system. (a) Neocortical lobes. (By Henry Vandyke Carter – Henry Gray (1918) *Anatomy of the Human Body*. Modifications: vectorization using CorelDraw. Public Domain, https://commons.wikimedia.org/w/index. php?curid=1676555.) (b) Two streams of information flow through neocortical visual processing hierarchy.

stellate neuron that was also activated axosomatically, and a layer 2/3 pyramidal neuron that was activated antidromically. In (c)-(d), neurons excited conventionally have solid axons, neurons excited antidromically have dotted axons ("u" – up-quark, "d" – down-quark, "e$^+$" – anti-electron, "$\overline{V}_e$" – electron-neutrino).

Figure 20: Hypothesized neuronal process analogous to Z^0 boson decay. (a) A layer 6 corticothalamic pyramidal neuron sends an axoaxonal projection up to layer 4 where it targets a lateral projection from one layer 4 stellate neuron to another. (b) The final products are one layer 4 stellate neuron activated axosomatically and the other antidromically. In (a)-(b), neurons excited conventionally have solid axons, neurons excited antidromically have dotted axons ("$\overline{V}_e$" – electron-neutrino, "$\overline{V}_e$" – anti-electron-neutrino).

Figure 21: Bifurcation diagram of the one-dimensional discrete logistics map.

Figure 22: Self-similarity in the period three window (r≈ 3.84) of the one-dimensional discrete logistics map bifurcation diagram.

Figure 23: The universe according to B-U IF—a system with the combined features of an M-theoretical Holographic Universe and the human thalamocortical-hippocampal system during a NREM spindle and sharp-wave/ripple event.

List of Tables

Table 1: The three generations of particles in the Standard Model of Particle Physics are like energy levels: the lowest energy level being the first generation (normal matter of the everyday universe), an intermediate energy level corresponding to the second generation, and the highest energy level corresponding to the third generation.

Table 2: Fundamental force particles. *The graviton is only theorized, i.e., unconfirmed, making gravity the lone known force that has yet to be fully incorporated into the Standard Model.

Table 3: Comparison of virtual and real photons.

Table 4: Quark, antiquark, and gluon color charge. Although the theory seems to allow for nine gluon color charges, detailed analysis of the theory—backed by observational evidence—reveals that there are in actuality only eight such charges.

Table 5a: Correspondences that define the isomorphism existing between our brains and the universe.

Table 5b: Correspondences (cont'd)

Table 5c: Correspondences (cont'd)

Table 5d: Correspondences (cont'd)

Preface

What this Book is About

In this text, I make the case that a physical model of the universe exists within the human brain. This model is not the same as the ones we are more familiar with, like the perceptual or higher-level cognitive ones that our brains create to best represent our immediate environment and situations, or the conceptual ones in the form of religions, philosophies, and sciences that we humans collectively create to best represent the universe and our place in it. Rather, this model arises due to the structural organization and dynamics of a particular subsystem in the brain, a subsystem that I assert most resembles our present day universe at a particular time during our nightly sleep cycles. This book, a product of eleven years of focused research, analyses, and writing, represents my best attempt to communicate this potentially worldview-transforming insight and the evidence to support it.

I will refer to the ideas that I present in these pages as the *Brain-Universe Isomorphism Framework (B-U IF)*. Its principal assertion is that at certain times, portions of the human brain define a system that is isomorphic to our current universe. This would mean that the universe has self-similar qualities and we, or at least portions of our brains, are the miniature copies of it that exist on a smaller spatiotemporal scale. In general, fractal phenomena are not at all uncommon because they can be found all throughout nature, even on the largest scales of the observed universe in the distribution of galaxies and galactic structures. However, given the current view of the universe emerging out of modern science, a fractal distribution of matter is just a subset of what it would mean for the universe to be self-similar. In other words, there are concepts being introduced by fields within physics, like string theory, that have a dramatic effect on

1

what a miniature copy of the universe would actually look like.

Who Should Care and Why?

Scientists and philosophers on the quest for a "theory-of-everything" will benefit from this text because if it is true that the human brain can be used as a physical model of the universe, then these researchers can use this principle as a guide, illuminating the way to the theory that they know in their hearts is somewhere out there, just waiting to be discovered by humanity. Thus far, scientists have been able to construct beautiful mathematical frameworks, one of which is string theory, that they believe gives them the best chance to describe reality on the most fundamental level, even more fundamental than our most cherished, well-established theories, such as quantum mechanics and relativity theory. However, and particularly in the case of string theory, these scientists do not possess the technological prowess to experimentally verify their theory's predictions; the consensus on this predicament is that, because experimentation and observation are crucial steps in the scientific method, string theory will not be fully accepted as a legitimate scientific theory that reflects reality as long as this lack of experimental verification persists. Therefore, a physical model upon which observations could be made would go a long way toward establishing string theory's relevance to physical reality.

In addition to professional scientists and philosophers on the hunt for a theory that is capable of explaining the physical universe, anyone else who enjoys pondering the nature of reality, such as those who can no longer ignore inconsistencies in their current worldview, may find that this text imparts profound insight because B-U IF provides answers to our most fundamental philosophical questions like: what is the nature of reality and god, and what is the purpose of life? In other words, B-U IF is a self-consistent worldview that extends the reach of our scientific knowledge to address such concepts as god, our relationship to the universe, and the meaning of existence. For those individuals—professionals and laymen alike—who wish to catch a glimpse of how humans stand in relation to the universe, a deep dive into descriptions of the universe and the brain emerging out of modern science is a requirement, an arduous yet most important and rewarding one.

The Motivation

When I think back, it seems as though I have always been interested in the nature of reality. While a child growing up in New Jersey, there came a time when I realized that, in a sense, both religion and science attempt to provide explanations for what reality is. I could also sense incompatibility between these systems-of-thought, so I would often wonder how to decide between the two. Both impinged upon me from the environment that I was subjected to—religion from family, and science from school, books, and TV. I felt that the truth was present somewhere within this cluttered signal. Pretty soon, however, it became obvious that there was more resonance between me and science than there was between me and the particular religion adhered to by my family, the Baptist sect of Christianity. This resonance has largely determined the trajectory that my mind has embarked on ever since I left NJ in pursuit of a higher learning.

First stop: Morehouse College in Atlanta, GA. It was here where I severed whatever little was remaining of the tether binding me to the religion that I had been exposed to up to that point in my life. I was aided in this process by being in the company of some of the brightest young African American male minds in the world (and female minds when Spelman College is included). It was an environment where independent thought, creativity, and bravery prevailed. By the end of my first semester at "The House", I decided to let what I felt were my finest qualities shine. Consequently, I decided to major in mathematics, a decision that proved pivotal because it was this classroom experience that is responsible for teaching me how to reason effectively and evaluate arguments.

It was also during my time at Morehouse that I became exposed to non-Western views of the world, such as those held by ancient Egyptians and many Asian cultures. I began reading about hermetic philosophy, a term used to describe a view of the world whose origins can be traced to ancient Egypt, and whose subsequent evolution was most notably influenced by ancient Greek translations of the older Egyptian teachings. I also began studying the metaphysical teachings of some Asian schools-of-thought, like Taoism, Buddhism, and Hinduism. What I learned from this experience is that I shouldn't be so quick to close the door on all religious teachings, that apparently, there is some resonance between me and some of these teachings, particularly the ones that more closely resemble pure metaphysical principles as opposed to the mythical dogma that can dominate religion.

I graduated from Morehouse in May of 2000, and enrolled at Hampton University (HU) in Hampton, VA, the alma mater of my sisters. Here, I began studying physics. I finished in 2003 with a Master of Science in physics, concentration in atmospheric science. The significance of my time at HU is that it is when I began to learn about the physical world, and how to conduct thorough research and communicate the results to the scientific community and general public.

During the tail end of my stay in Hampton, I became interested in a potential link between modern physics and some of the esoteric teachings of ancient Egyptian and Asian schools-of-thought. To my surprise, this time period—spring of 2003—just so happened to be a period of renewed interest in the similarities that exist between Eastern thought and quantum mechanics, one of the crown jewels of modern physics. There were numerous books and magazine articles published on the topic, and perhaps the most discussed aspect of this debate involved the role of consciousness in the universe. Eastern philosophies teach that consciousness is the fundamental basis of reality, and many scientists and philosophers believed that conscious observers play a crucial role in the quantum processes that occur in the universe. Those who held the latter view argued that physical reality only exists in the presence of conscious observers because it is their consciousness that plays a causal role in the manifestation of physical reality from the unphysical quantum possibilities. If this is the case, it would be consistent with the Eastern teaching that consciousness is the root of all reality. However, the excitement about a possible underlying connection between these esoteric teachings and quantum mechanics seemed to die down midway through the first decade of this century, when quantum physicists began to increasingly interpret their findings within the context of information transfer. This way of viewing quantum processes removed the seemingly special role of the conscious observer in the outcome of quantum experiments, and I feel that this ultimately took the wind out of the sails of this most recent mainstream movement to show similarity between the teachings of esoteric traditions and modern physics.

After graduating from HU, I became a physicist at a US national laboratory where my experiences have reinforced my standards of research and ability to communicate my work. Even while settling into this career, I never lost sight of a potential connection between the teachings of esoteric philosophies and modern physics, and after some time, I realized that there are common themes in the esoteric teachings that have yet to be fully explored, namely, that the universe is created mentally and that we are the microcosm of it and/or the entity that creates it. I took from this that it might be possible to gain useful insight about the nature of the universe by comparing its structural organization and dynamics to that of the brain. Contrast this with the emphasis of the esoteric teachings on instructing those who wish to comprehend the nature of reality to first become familiar with their mind via altered states of consciousness, such as meditation and lucid dreaming; this is a subjective approach to comprehending the nature of reality. I, on the other hand, decided to take a much more objective approach. I saw this as an opportunity to conduct valuable philosophical and scientific research within the domain of my true passion— using the systems-of-thought that I naturally gravitate to, esoteric philosophy and modern science, to comprehend the nature of reality. Therefore, inspired by my interpretation of the instructions laid out by the esoteric teachings, I set out to compare what some of the leading scientific theories have to say about the universe and the brain. During my research, I have come across other reasons, ones offered up by modern science, why the universe and brain should be systematically compared. In other words, a plausible scientific argument can be made that there may be a deep fundamental similarity between the universe and our brains that makes no reference to the esoteric philosophical teachings that I have just mentioned. I present this argument in the introductory chapter of this book.

In his seminal text *The Tao of Physics: An Exploration of the Parallels between Modern Physics and Eastern Mysticism* (first published in 1975),[1] Fritjof Capra does an excellent job presenting a myriad of parallels that exist between the respective worldviews of various Eastern schools-of-thought and modern science. However, he stops just shy of the critical insight that I consider in this text, insight that I gleaned from my own analysis of the teachings of esoteric philosophy, which in my usage of the term includes Eastern thought as well as hermetic philosophical teachings. Again, this critical insight is that we should perform a comparative analysis of the structural organization and dynamics of the universe and the brain. I am unaware of analyses like this other than the research of

Michael Talbot who in his 1991 book *The Holographic Universe*[2] presented parallels between the respective teachings of the physicist David Bohm and the neuroscientist Karl Pibram. More recently, however, there are scientists and philosophers who have proposed views of the universe similar to the views expressed here in this book. Robert Lanza and Bob Berman state in their 2009 book *Biocentrism: How Life and Consciousness are the Keys to Understanding the True Nature of the Universe*[3] that the neuronal circuitry in our brains "contains" the logic of space and time and that physicists attempting to understand reality would probably benefit from also considering the insights gained from studying the brain. In addition, Bernardo Kastrup suggests in his book *The Idea of the World: A multi-disciplinary argument for the mental nature of reality*[4] that the universe has brain-like structural organization and dynamics but that it is not necessarily analogous to human brains. I, in this current text, extend in both depth and scope the systematic brain-universe comparison hinted at and argued for by these great researchers and authors.

The Method

Before a comparison can be carried out, it is necessary that I clearly identify and define the models that I use to represent the universe and the brain because in the fields of physics and neuroscience, there are numerous competing theories. To be clear, the qualitative conceptual models that I present in this book are just two of many descriptions of "the way things could be" emerging out of their respective disciplines—physics and neuroscience. While it is true that I settled on these particular models over the many other possibilities because I found the interesting result that they are very similar to each other, I firmly assert that the two models are front runners in their respective disciplines. If it ultimately turns out that, in fact, there are very similar views of the universe and the brain that are based on leading theories in physics and neuroscience, respectively, then it will be up to the scientific community to determine if it is just a coincidence or an indication that there is a deep connection between the structural organization and dynamics of the two systems.

On the smallest and largest spatiotemporal scales, the model that I present to represent the universe is fundamentally based on string theory, but physical processes on intervening spatiotemporal scales are described by quantum physics, classical physics, relativistic physics, cosmology, etc. Furthermore, because the universe has been highly dynamic since the start, "big history" is another discipline that I found useful for capturing universal qualities. The model that I define to represent the universe is presented in Part I. To construct a qualitative conceptual model of the brain, which is presented in Part II, I incorporate many well-established findings and leading theoretical proposals within neuroscience. The overarching framework that I use to describe higher-order brain organization and function is based on Christof Koch's ideas on the intermediate-level theory of consciousness that he presented in his book *The Quest for Consciousness*.[5] Strategically along the way in Part II, I install "waymarkers" in the form of specified correspondences between the content of Parts I and II—the purpose of these waymarkers are to prime the reader on how to see correspondence between the universe and brain.

Research on the universe and the brain has revealed that both systems are highly complex, and that a fundamental activity of both is to process information; therefore, I have also consulted complexity theory and information theory, something that both physicists and neuroscientists have begun to do as well. And thanks to the highly interdisciplinary nature of this research, I also found general systems theory to be useful because it provides insights that pertain to systems of all types, no matter what their fundamental constituents are.

To build up my qualitative models of the universe and brain, I relied heavily on work published by other scientists and writers, some of whom are working scientists within the many subdisciplines relevant to this research, and some are science journalists who cover the work of working scientists. Where necessary, I provide references in support of what I claim. My hope is that the addition of this information will facilitate fact-checking of this work, something that I highly encourage people to do. Along with citation numbers appearing within the text, corresponding to reference sources listed in References, I also tried to provide page numbers as much as possible, whenever it was practical to do so, to more precisely support what is currently being discussed. It is my wish that this effort to preserve transparency serves to enrich the experience of interested readers.

With this research, I have integrated knowledge that has been uncovered by modern science into as complete and consistent a description of the universe and the brain that I possibly could. But what I do not do in this book is cover the story behind how each individual discovery in the various scientific disciplines were made throughout the years—my focus is almost exclusively on defining the conceptual models that I use to represent the universe and the brain. I have used the best available information to define these models and I present some equivalence between the two in Part III. It's important for me to emphasize here that I do not believe the examples of equivalence, or, correspondences that I provide in Part III are exhaustive by any means. In defining models for the universe and brain, I went into a fair amount of detail, including as much as I felt I had a comfortable understanding of, and whichever details are not included in the list of correspondences in Part III can serve as clues to other scientists and metaphysicists who wish to pick up where I left off with the research and expand upon the details of the isomorphism. Just like when examining a fractal, if the reader looks deep enough, and long enough, he or she is sure to see more.

Ultimately, this research has allowed me to construct a self-consistent, credible, and powerful worldview, one that I believe will be a useful concept to humanity as we continue on our trajectory through universal evolution. This is a bold statement for sure, and one question that may be on the mind of many readers is: if it is true that the structural organization and dynamics of the human brain and the universe are identical, then why haven't we noticed this already? For one, I have found that string theory offers up a theoretical framework for modeling the universe in such a way that the parallel between the universe and the brain can be seen in great detail, but note that some of the most pivotal insights to emerge from string theory did not do so until the late 1990s. In addition, many critical neuroscientific insights that allow a glimpse at the correspondence between the universe and brain in any significant detail also did not emerge until late last millennium and continue to emerge at a rapid pace. This leaves about a 20-year window—roughly 2000 to today (2020)—when it would have been possible to sift through the massive amount of information necessary to assess a potential correspondence between the universe and brain. But I would also argue that 2000 is still relatively early to expect that someone would perform this type of analysis because the insight within the respective fields of string theory and neuroscience most likely existed within local pockets of academia, expressed in very technical terms in obscure journals and not well suited for a broader audience. It would take at least another few years before the great work of scientists and science writers would make these insights digestible for non-experts. I began researching both string theory and neuroscience in 2009, right around the time it was just becoming possible to even detect the type of high resolution correspondence that I present for you in this text.

Chapter 1

Introduction

The Good Regulator Theorem in Light of Recent Findings in Neuroscience

Consider for a moment a seminal paper written by Roger C. Conant and W. Ross Ashby.[6] In that paper the authors introduced the cybernetic insight that "every good regulator of a system must do so by forming models of that system." When Conant and Ashby considered the extreme case of this theorem, one where the regulator is most optimum, they concluded that "... any regulator that is maximally both successful and simple must be isomorphic with the system being regulated." By isomorphic, the authors meant that it is possible through some type of transformation, or mapping, to view the regulatory system as having the same structural organization and dynamics as the system being regulated.

Conant and Ashby also suggested that when we consider humans in light of this theorem, it becomes apparent that "... the living brain, so far as it is to be successful and efficient as a regulator for survival, must proceed, in learning, by the formation of a model (or models) of its environment". In this corollary the authors explicitly acknowledge the possibility that the human brain can form multiple models to represent various aspects of the environment. This is a point that modern neuroscience is indeed finding to be true.[7]

But is the human brain more than just "successful and efficient" as a regulator for our survival? In other words, could the human brain be more like the most optimum regulator considered by Conant and Ashby, one that is "maximally both successful and efficient"? Interestingly, recent research into this matter has found that the brains of humans (and of other animals often studied to gain insight into the human brain) may indeed possess optimum, or at least near optimum, qualities when it comes to certain types of functions and structural organization. Consider the following:

There is evidence of optimal learning.

- Scientists have developed a mathematical model of optimal learning referred to as the "ideal observer", and it has been shown that during a learning task, humans can perform just as good as the ideal observer and they do so by evaluating the reliability of what gets learned and the confidence levels that they have in their predictions just like the ideal observer model does.[8]
- Humans and rats have shown the ability to optimally accumulate evidence for decision-making.[9]

There is evidence of optimal perception.

- Scientists have shown that when a person simultaneously inspects an object with their hands and eyes, the brain combines both forms of perceptual information in an optimal fashion.[10]
- Processes in macaque monkey brains that coordinate saccadic eye movements and attention do so in a way that visual stimuli can be optimally tracked and processed.[11]

There is evidence of optimal navigation.

- An artificial intelligence (AI) program, inspired by the human brain, has been designed to, over time, optimize navigation within challenging, unfamiliar, and changing environments. In so doing, the AI system has shown the ability to recreate the same type of signature electrical activity displayed by neurons in the human brain that specialize in navigation, suggesting that these neurons may also employ an optimal algorithm to carry out their function.[12]

There is evidence of near-optimum structural organization.

- Structural networks in the human cortex have 89% of the connections that a highly idealized model of the cortex has, one that optimizes the transfer of information.[13]

It's important to note here that the use of the word "optimum" to describe function in the human brain does not imply perfection, as if the human brain is incapable of error and always has complete knowledge of every situation. The ideal observer model produces the best results at performing a particular task that any physical system can possibly achieve given the information that is available. If perfect knowledge is made available to the ideal observer, it will perform perfectly. However, when there is uncertainty, the ideal observer will make errors.[14] Therefore, even though the human brain has shown the ability to perform as well as the ideal observer model for some cognitive functions, it too will make errors under conditions of uncertainty. Since there is an irreducible amount of uncertainty in the world due to such things as quantum uncertainty on the most fundamental level, deterministic chaos on the classical level, and the finite nature of human experience in both space and time, it is expected that the human brain will always be prone to errors.

A similar situation could exist on the collective level as well. That is, if each of us can perform as well as the ideal observer model as we live and we learn, then it's also possible that we learn optimally on a collective level, meaning that the acquisition of humanity's knowledge over the course of time could be occurring in an optimal fashion. Just like on the individual level, however, fundamental amounts of uncertainty will again act as a constraint on just how well humanity can collectively process information.

What Constitutes Humanity's Environment?

Together, the Conant-Ashby Theorem and the recent findings of optimal organization and function in the brain suggest that on a fundamental level, the structural organization and dynamics of our brain and our environment may be very similar in some way. Before this inference can be fully investigated, a working definition of "our environment" will be necessary. In the broadest sense, the environment can be defined as ALL of the factors that CAN act on us, whether through our five senses or our scientific equipment, and influence our physical states or actions. Based on this definition, our environment is a concept that is much more expansive than simply the things that we can observe with our eyes, hear with our ears, smell with our nose, feel with our skin, or taste with our tongue. It's even more expansive than the concept of the environment implied by the phrase "environmental protection". This definition of environment includes things that may in fact influence our physical states or actions but yet still await discovery. Furthermore, there are things in our environment that we know exist but at any given moment we may not even perceive, or may not even be conscious of. However, the mere fact that we know a thing to exist means that it has affected our mental states and can help to shape our concept of the universe. Based on these considerations, the environment of humanity must *at least* be taken to be the entire *observed* universe, everything from fundamental particles that are on the smallest spatiotemporal scales observed via our most powerful particle accelerators, to the large-scale structure of the universe that occupies the largest spatiotemporal scales of observation allowed by our most sensitive telescopes. Put simply, I argue that if humanity can detect a phenomenon so that we can confirm that it exists, then it can be considered to be a part of our environment.

Academic disciplines within science and the humanities do a good job of explaining most of what we know about the observed universe. However, our scientific disciplines that attempt to describe the fundamental nature of the physical universe, such as classical mechanics, relativity theory, quantum mechanics, cosmology, etc., often predict phenomena that have yet to be observed and confirmed, like fundamental particles that are potential candidates to join the Standard Model of Particle Physics. Over time, some of these theoretical predictions may eventually be confirmed through observation and make the transition from purely theoretical phenomena to acknowledged aspects of the universe. One recent example of this type of transition is the Higgs boson, which was discovered at the Large Hadron Collider in 2012[15] after being first theorized in the 1960s. In addition to well-established fields within physics, there are emerging fields, like string theory, that attempt to extend upon the explanatory powers of our current theories. New theoretical proposals like these are ripe with entities that are currently of a strict theoretical nature but may one day be discovered and become an accepted feature of our universal environment.

In short, what we consider to constitute the universe—our environment—has changed over time as discoveries have been made and theoretical predictions have been verified. In fact, this process will continue as long as science continues to evolve and uncover fundamental aspects of the universe. Therefore, even the theoretical proposals that exist at or beyond the boundary of what we can currently observe play an important role in defining the different pictures of the universe that are emerging out of modern science.

Reframing the Good Regulator Theorem

The recent observations of optimality in the human brain suggest that our collective modeling of our environment, and ultimately of the universe, may occur in an optimal fashion, but does optimal modeling of the universe equate to optimal regulation? Humanity's ability to regulate anything in the universe to any extent is an attribute that continuously changes over time as our collective knowledge and technology evolve. We use what we learn about the universe to harness and manipulate the available energy and natural resources, making it possible to build the infrastructure of society, manufacture products, and perform other tasks necessary to fulfill our needs and desires. However, currently, and for the foreseeable future, it would appear that humans are not maximally successful and simple regulators of the universe simply because there are many phenomena that we are unable to regulate. As is often the case though, we can have detailed knowledge of the underlying physical processes taking place within a system, yet still lack the means, or even sometimes the desire, to implement this knowledge into a practical technological application that would allow us to exert some degree of regulation over the system. This is a case of our knowledge of the inner workings of a physical system outpacing our ability to exploit the knowledge in the form of a practical application.

Because it cannot be said that humanity regulates the universe in any global, or complete way, we cannot simply insert the brain as the regulator, and the universe as the regulated system, referred to in the Good Regulator Theorem as such:

> If the brain *is maximally both successful and simple as a regulator* of the universe, then it must be isomorphic with the universe.

However, the close relationship between regulation and modeling established by Conant and Ashby suggests that we can reframe the Good Regulator Theorem and consider the following:

Is it also the case that if the brain *is maximally both successful and simple as a modeler* of the universe, then it must be isomorphic with the universe?

The Good Regulator Theorem says that modeling is a mandatory step toward the achievement of successful and efficient regulation. Humanity's collective acquisition of knowledge has led to modern science, which produces a wide array of oftentimes highly detailed models of a multitude of phenomena in the universe. Humanity is indeed modeling the universe and may be doing so optimally, which could be evidenced by the fact that many scientists have concluded that we are close to the development of a "theory of everything" that successfully unifies quantum mechanics and general relativity to explain all known fundamental physical processes in the universe.[16] The recent observations of optimality in the human brain lead to the possibility that there is a deep similarity between the structural organization and dynamics of the universe and our brains, providing a potential clue as to how we can advance on our quest for a more thorough understanding of the universe.

Documented Evidence of Structural and Dynamical Similarities Between the Universe and Brain

Some scientists who study the universe and the brain have already begun to acknowledge that there are very important similarities between the respective systems that they study. For example, quantum theory says that on a fundamental level, spacetime is not continuous; rather, it is discrete, consisting of individual units on the smallest possible spatiotemporal scales. Furthermore, all physical events can be considered to be some form of interaction between entities at these discrete units of spacetime. Based on this formulation, the large-scale structure of spacetime in our universe can be viewed as a complex causal network where the discrete units of spacetime are nodes and the interactions between them are links, or, edges. It has been shown that the network encoding the large-scale structure and dynamics of spacetime is in many ways similar to the large-scale structure and dynamics of other complex networks that have appeared in nature and society.[17] The outermost part of the human brain, the neocortex, which is the most recent part the brain to have evolved and the part of the brain that is primarily responsible for our higher-order cognitive abilities, is one such complex network that has been identified as sharing structural and dynamical properties with this quantum causal network model of spacetime. It should be noted, however, that the researchers responsible for this discovery caution against interpreting this result to mean that the universe and the neocortex are identical.

In addition, there is now a trend in psychology to use a quantum mechanical framework to more accurately model aspects of human cognition, such as decision-making, memory, reasoning, and perception.[18,19,20] That is, rather than modeling decision-making using classical probability theory, psychologists are starting to employ concepts from quantum mechanics, such as indefinite superpositions of states and measurement-induced collapse of these states, to more accurately describe how the human brain functions. It should also be noted here that the scientists at the forefront of this trend do not claim that the brain functions as a quantum computer. Rather, what they have shown is that they can better capture the ability of our brains to function in uncertain and ambiguous situations by applying the mathematics of quantum mechanics to human cognition.

These two examples hint at how the universe and our brains could share similar structural organization and dynamics. In particular, these two examples compare our quantum description of the universe with certain aspects of brain structure and function. In this book, I present a more complete comparison of the two systems defined by the universe and the brain by also uncovering similarities between the brain and our other descriptions of the universe, such as classical mechanics, relativity theory, cosmology, string theory, etc.

What would be the Significance if the Human Brain is a Model of the Universe?

If it turns out that the human brain and the universe have analogous structural organization and dynamics, then it means that the brain can be used as a model of the universe, and vice versa. Such a principle can provide physicists searching for a "theory of everything" with a model system to make observations on, opening up a whole new way to investigate and validate their theories, a luxury that many of our most recent theoretical endeavors severely lack. For example, string theory is widely identified as one of the most promising frameworks for extending our understanding of the universe beyond the capabilities of relativity theory and quantum theory alone.[21] However, string theory predicts entities, such as strings and the tiny curled up extra dimensions that strings need to vibrate into, that have virtually no chance of ever being experimentally verified via conventional means, such as particle accelerators.[22] Therefore, the ability to make observations on a physical system that has properties very similar to those possessed by the universe will allow string theorists to establish, via observational evidence, string theory's relevance for describing the universe, an accomplishment that could potentially usher in a "third string theory revolution" — see "The Elegant Universe"[23] for more on string theory revolutions.

But note that if the human brain and the universe have analogous structural organization and dynamics, then ultimately the flow of information can occur in two directions. That is, it wouldn't just be physicists who would benefit from being bestowed a model system upon which they could make observations. Neuroscientists would also benefit because now a larger set of insights and concepts that comprise theoretical physics can become an inspiration and a guide to their future research efforts. Currently, there are numerous collaborations between neuroscientists and physicists where the toolbox used by physicists to characterize physical systems has experienced great success when applied to the study of the brain.[24] However, identifying a picture of the universe emerging out of physics as a viable model of the human brain would provide higher-level intuition and theoretical constraints for scientists who attempt to decode the ways of the brain.

How Self-similar can the Universe be?

If the human brain is a model of the universe, it would mean that the universe has self-similar qualities like a fractal and that we, or at least portions of our brains, are the miniature versions of it that exist on a smaller spatiotemporal scale. Fractal phenomena like this are not at all uncommon. They can be found all throughout nature due to the prevalence of nonlinear dynamical processes.[25(pp175-176)] However, it should be noted that if indeed the universe is fractal, then based on the known properties of fractals, the nature of any miniature copies would depend on the type of self-similarity supported by the universe.[26(p63)] In other words, if the universe is like an idealized and infinite mathematical fractal, then in theory, it is indeed possible for strict self-similarity—exact replicas—to exist on smaller spatiotemporal scales, but if the universe is a finite fractal that only supports approximate self-similarity, then it is only possible for an approximate replica system to exist inside it.

As far as we know, the only ideal mathematical fractals that exist do so in our computer simulations. What we have seen in nature are fractal structures and processes that exhibit approximate self-similarity.[26(p63)] In well-studied nonlinear dynamical systems that exhibit self-similarity (like the Mandelbrot set and logistics map), what often exists is a multitude of substructures and processes that range anywhere from subtly similar to nearly exact copies of the larger system. However, in many cases, even though the miniature copies in these structures are not exact copies of the larger system, they are similar enough that they can still be used as a model of the larger system. This may also be the case for the human brain in the event that the universe is self-similar and the human brain is a manifestation of this self-similarity. In other words, the similarity doesn't have to necessarily be exact, but it could be so close that knowledge of the analogous structural organization and dynamics may prove to be a useful principle for advancing our understanding of both the universe and the brain.

The high likelihood that no two human brains can ever be exactly the same implies that what may be isomorphic to the universe is a general topological structure common to all human brains, yet still distinguishable from the rest of the animal kingdom. It may be possible to identify a human-specific topology through the use of algebraic topology, a mathematical discipline that is increasingly applied to the study of the brain on multiple spatial scales, revealing structures whose features can be readily characterized using well-known topological concepts, such as simplices, simplicial complexes, and cavities.[27,28,29] Interestingly, these very same, or other very similar, analysis tools have also proven effective in our attempts to improve and extend our ability to describe the universe on the most fundamental of levels. Algebraic topology is currently being used to encode the probabilities of the various particle interactions allowed by the Standard Model of Particle Physics into an abstract topological object known as the "amplituhedron" that can be characterized using the same topological concepts used to characterize network structures in the brain, namely, simplices, simplicial complexes, and cavities, etc.[30,31] Likewise, the mathematics of string theory is based on algebraic geometry, a very high degree of functionality for modeling the universe. If the human brain is a physical model of the universe, then the Good Regulator Theorem suggests that it's a case of form defining function. Furthermore, I speculate that consciousness itself may even be a product of the universe's ability to become self-similar, such as when evolution produces species of life on Earth whose brains have structural and dynamical similarities to the universe as a whole. The stricter the similarity between a brain and the universe, the richer the conscious experience that would be possible within that brain.

Conclusion

By reframing the Conant-Ashby Theorem and considering the implications of recent findings in neuroscience about optimal function and organization in the human brain, I argue that the human brain and humanity's environment, which I take ultimately to be the entire universe, should be compared to determine if they have similar structural organization and dynamics. In the chapters that follow, I present a systematic comparison of the ideas emerging out of physics concerning the fundamental nature of the universe, and the ideas emerging out of neuroscience concerning the fundamental nature of the brain and mind. I identify analogous structural organization and dynamics of these two systems that can serve as a powerful guiding principle for the advancement of our understanding of the universe, the human brain, and the relationship between the two.

Part I

Universe

Chapter 2

Standard Model of Particle Physics

Systems of the Universe

A system is defined as a collection of parts that come together through some form of interaction to form a single functional entity. The universe itself is simultaneously one single system and a collection of interacting subsystems that can be categorized into hierarchies based on their spatial and temporal—spatiotemporal—extent and levels of complexity[37(pp51-54),38] (Figure 1). The spatiotemporal and complexity continuums encompassing all systems in the universe vary widely and those systems that occupy different positions within these continuums exist alongside each other and are able to interact to varying extents—all the way from the most elementary unit to the entire universe itself. These systems tend to be both parts

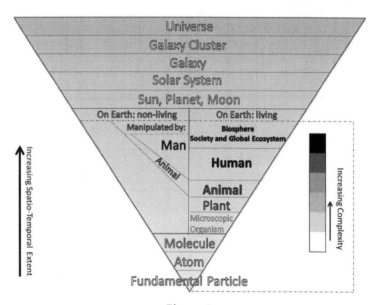

Figure 1

27

and wholes at the same time, meaning that on one level of the spatiotemporal hierarchy, they are a single functional entity made from a collection of parts existing on levels below, while on larger scales, they become the sub-parts of larger systems. In general, the interaction of systems on lower levels creates systems existing on higher levels that have emergent properties, or, properties possessed by the system as a whole but not by the individual parts. Ultimately, it is the interaction of the most fundamental units and systems, those existing on the lowest level of the spatiotemporal hierarchy, that create the wide variety of matter that constitutes the universe.

Fundamental Particles

The Standard Model of Particle Physics, or Standard Model for short, describes the most fundamental particles existing on the lowest level of the spatiotemporal-complexity hierarchy in Figure 1 and the ways in which they interact to form all matter in the universe. It says that all physical systems consist of matter particles (parts) and force particles (the means by which the parts interact).[39(p352)] Most people are by now familiar with the concept that all matter in the universe is made out of atoms and that atoms themselves consist of a nucleus and some number of electrons. The electron is considered a fundamental matter particle because it is structureless, meaning, it is not believed to be composed of a system of more elementary particles.[40(p5)] The nucleus on the other hand is made from more elementary units called protons and neutrons and it turns out that these particles consist of particles even more elementary called quarks. Coming in two varieties, "up" and "down", these quarks are structureless fundamental matter particles like the electron, making both types of particle the basic building blocks of normal matter in the universe.[39(p262),40(pp5,7)] There does, however, exist another fundamental matter particle, the neutrino$_e$ (referred to as an electron-neutrino). This is a particle that is closely associated with the electron and plays a role in the process of up- and down-quarks changing into each other.[40(pp62-63)]

Parts

The Standard Model says that there are three such groups, or "generations", of fundamental matter particles organized according to their masses and the energies required for their manifestation, and that each generation is identical in every other way such as their interaction via the fundamental forces.[40(p133),41] The first generation of fundamental matter particles consists of the particles discussed above, namely, the electron, neutrino$_e$, up-quark, and down-quark; today, normal matter is predominantly made up of these particles thanks to the relatively low energy densities existing throughout the universe.[39(p357)] Because of their ability to probe the depths of matter at extremely high energies, modern particle accelerators have led to the discovery of the other two generations of fundamental matter particles.[40(p125),41] The heavier, and hence, higher-generation particles appear early on in the moments after high energy particle collisions because they require the highest energy densities for their creation, similar to the densities that existed just after the Big Bang. These particles tend to be unstable and disintegrate quickly into lighter particles whose tracks are picked up by the collider's detectors.[21(p95)] This decay usually continues until the particles of the first generation are produced, the lightest and most familiar species in the Standard Model.[39(p357),42] The particles in the second generation are the muon, neutrino$_\mu$ (muon-neutrino), strange-quark, and charm-quark; while the particles in the third generation are the tau, neutrino$_\tau$ (tau-neutrino), top-quark, and bottom-quark. There is no good explanation as to why there are these three generations of particles;[40(p133)] nevertheless, it is the interaction of these matter particles via the exchange of force particles that results in all the normal matter of the universe (Table 1).

Generation	Leptons			Quarks		
	particle	mass (GeV/c²)	charge	particle	mass (GeV/c²)	charge
1	electron neutrino	< 1×10⁸	0	up	0.003	⅔
	electron	0.000511	-1	down	0.006	-⅓
2	muon neutrino	< 0.0002	0	strange	0.1	-⅓
	muon	0.106	-1	charm	1.3	⅔
3	tau neutrino	< 0.02	0	bottom	4.3	-⅓
	tau	1.7771	-1	top	175	⅔

Table 1

For every particle in Table 1, and the larger particles that they can form, there is what's called an antiparticle. These antiparticles have all the same characteristics as the regular particles except they have the opposite charge, and in a sense, may even be viewed as the regular particle moving backwards in time. Collisions between matter and the corresponding antimatter result in the complete annihilation of both, leaving behind pure energy.[40(p125)] On the other hand, when high energy collisions between two normal matter particles occur inside accelerators, the energy released from the collision decays into equal amounts of matter and antimatter. In today's universe, however, long after the antiparticle-producing, high energy conditions of the Big Bang have subsided, there is a large asymmetry between matter and antimatter[43] — the universe consists of much more matter than antimatter. Physicists are still trying to ascertain what it is about the fundamental nature of the universe that creates this imbalance.[40(p51)]

Interactions

For each fundamental matter particle in the three generations, the interactions between them are facilitated by some combination of the four fundamental forces: gravity, electromagnetism, strong nuclear, and weak nuclear. The Standard Model says that these forces are mediated by the exchange of fundamental force particles between the matter particles.[16] For the gravitational, electromagnetic, strong nuclear, and weak nuclear forces, these particles are the graviton (which is still only theorized and not yet officially included in the Standard Model), photon, gluon, and $W^{\pm}$ and Z^0 particles, respectively (Table 2). These force particles are a part of another category of particles called bosons, which differ from the matter particles in that more than one boson of any kind can simultaneously occupy the same physical state, a situation that is impossible for the fundamental matter particles. The Higgs boson, a particle responsible for giving each fundamental particle the amount of mass that it possesses,[44] has just joined the list of known bosons since it was recently observed inside the Large Hadron Collider (LHC).

Force	Particle	Mass (GeV/c²)	Charge
gravity*	graviton	0	0
electromagnetic	photon	0	0
strong	gluon	0	0
weak	W⁻ boson	80.4	-1
	W⁺ boson	80.4	+1
	Z⁰ boson	91.187	0

Table 2

Gravitational Interactions

The particle theorized to mediate the force of gravity is the graviton.[16,39(p255)] This massless particle with zero charge is believed to be emitted out at the speed of light into the environment by all particles and particle agglomerates with mass, giving them a gravitational field that is infinite in extent, although the field's strength becomes negligible at distances very far from the source. Gravity is an attractive force that acts on all matter and light; the more massive the source object, the stronger its gravitational influence on matter and light in its surroundings. The gravitational force is negligible on the tiny scales of fundamental particles and only becomes significant when physical systems much larger than an atom are involved.[40(p47)] I cover the force of gravity more thoroughly when discussing Relativity Theory in Chapter 4.

Electromagnetic Interactions

All particles with charge can interact by exchanging photons, the mediators of the electromagnetic force. The two types of charge that exist are called positive and negative, and the degree to which an object is positively or negatively charged is determined by the difference in the number of positively and negatively charged particles comprising it; the larger the charge, the stronger the object's electromagnetic influence on charged objects in its environment. Protons are used to set the standard magnitude of charge, having a positive charge that has been set equal to unity, while the charge of an electron is -1, equal and opposite to that of the proton. The interaction between positive and negative charge is such that like charges repel and opposites attract. When objects with opposite charges of equal strength come together to form a single system, their charges counterbalance each other so that the system as a whole has a net charge equal to zero. In this way, the electromagnetic force can be looked at as one that a charged system uses in an attempt to become an uncharged system.

To understand how two charged systems interact via the electromagnetic force, it is necessary to consider systems whose charge configuration is constant (static) over the period of observation and systems whose charge configuration is changing (accelerating) because the nature of the photons involved in these two scenarios is different (Table 3). For charged systems whose charge configuration is unchanging, the photons that mediate the resulting static electric force are only *theorized* and never *directly* observed. In other words, these photons are not detected as particles, rather, it is their effect on charged matter in the environment that is observed. The photons in this case are called "virtual" photons.

On the other hand, systems whose charge configuration

changes in time produce real photons that are detectable as particles because they transfer momentum and energy to charged fundamental particles. Unlike the virtual photons, real photons do not exert an attractive or repulsive force that depends on the charges of the emitting and receiving particles. Nevertheless, these real photons—the ones that are not just theorized but also detectable as particles—are associated with visible light and the rest of the electromagnetic spectrum. And just like gravity, the electromagnetic field of a particle is transmitted at the speed of light and has infinite extent, although the strength of the field very far from the source is negligible.

Photons manifest in two ways		
	virtual photon	real photon
force	electrostatic	electrodynamic
detectable as particle	no	yes
effect on charged particles	opposite charges attract, like repel	transfers energy

Table 3

Strong Nuclear Interactions

The strong nuclear force results when quarks interact by exchanging gluons, causing them to cluster and form larger particles, an example being the particles comprising the nucleus of atoms—protons and neutrons. It is the force with the shortest range, operating only over a distance the size of a nucleon; in the case of the proton, two up-quarks and one down-quark exchange gluons to form a single system while in the neutron, it is one up-quark and two down-quarks that interact.

Quarks and gluons have a property called color charge that determines their strong nuclear interactions. It should be mentioned that the use of the term color to refer to this property was never meant to indicate that quarks are actually colored in the sense that we experience color through visual means; it is just a conceptual tool used to describe a property that, through experiment and theory, quarks are known to possess. While quarks and gluons themselves have this color charge, the particle clusters that they form do not. One of the ways that uncolored systems form is when three quarks, each with a different color charge, come together to form protons and neutrons. On the other hand, a quark and an antiquark can also cluster to form an uncolored system. If red, blue, and green are used to represent the three possible quark color charges, then the opposite color charges of antiquarks are represented by anti-red, anti-blue, and anti-green. Like-colored quarks close enough are repelled, opposite-colored quarks experience strong attractions, and quarks of different yet not opposite color experience a strong but weaker attraction. The strength of the strong nuclear force also depends on the distance separating the quarks—the farther apart they are while still within the diameter of a proton, the stronger they are attracted, but the closer they are, the weaker the attraction. This ensures that the quarks and gluons remain in tight clusters.[42] In fact, because of the strong force, quarks and gluons have never been observed free—usually only in clusters of two or three.[40(pp97-98)] Furthermore, the strong nuclear force acts independent of the electrical charges of the particles involved, i.e., the electrical charges of the quarks have no effect on the strength of the strong nuclear force.[40(p109)]

The color charge of gluons is a bit more complicated than the color charge of quarks because they are a combination of the colors allowed for quarks and those allowed for antiquarks (Table 4). For example, a gluon could have the color charge that is red and anti-blue, or green and anti-red; the combination of the two color charges, one for quark and one for antiquark, constitutes a single color charge for gluons. Gluon color charges are indicative of the communication taking place between the quarks forming clusters, namely, it comes from the color of the quark emitting the gluon and the anti-color of the quark receiving the gluon. The result of quarks communicating in this way is that they swap color charges every time they engage in strong nuclear interactions and they do so in such a way that the particle their clustering forms, such as a proton or a neutron, is always uncolored.

Particle	Color charge
quark	red (r), blue (b), green (g)
antiquark	anti-red ($\bar{r}$), anti-blue ($\bar{b}$), anti-green ($\bar{g}$)
gluon	$r\bar{r}, r\bar{b}, r\bar{g}$ $b\bar{r}, b\bar{b}, b\bar{g}$ $g\bar{r}, g\bar{b}, g\bar{g}$

Table 4

In many ways, color interactions between quarks are analogous to the interactions among electric charges, except that it occurs in threes as opposed to twos.[40(pp93,97)] In the analogy, color charge generates the strong nuclear force in the same way that electric charge generates electromagnetic forces; massless gluons are the particles that mediate the strong force while the massless photon mediates the electromagnetic force; the acceleration of color radiates gluons while the acceleration of charge radiates photons.[40(pp97-99)] The effect of both forces on matter is to create systems whose net charge, color or electric, is zero. In addition, there appears to be a connection between the tendency of quarks to cluster in threes and the fact that they have third-fractional electric charges, 2/3 for the up-quark and -1/3 for the down-quark. When they cluster to form protons and neutrons, they form a system with whole number charge, 1 and 0, respectively.[40(pp174-175)] Protons and neutrons also turn out to be uncolored systems because as we saw earlier, when three quarks with different colors combine to form one system, the result is a system without color charge. For this reason, protons and neutrons do not exchange gluons the way that quarks do. Moreover, electrons and neutrinos (in general, all leptons in Table 1), which both have whole number charges, also do not possess color and, therefore, do not experience strong interactions.[40(p97)]

Weak Nuclear Interactions

The particles that mediate the weak nuclear force are the $W^{\pm}$ and Z^0 bosons—the superscripts indicate the electric charge of the particles. The weak force enables quarks within the same generation to change into each other, a process known as beta decay which also involves the leptons of the generation.[40(p154)] For example, when a neutron transmutes into a proton, one of its down-quarks turns into an up-quark and in the process, emits a W^- boson which quickly decays into an anti-neutrino$_e$ and an electron (Figure 2). Weak interactions such as this can occur to achieve a state of stability within an atom, something that could occur when the mismatch between the numbers of protons and neutrons within the nucleus becomes too large.[40(pp37,42)] A similar interaction exists for the second and third generations; for example, a charm-quark can turn into a strange-quark through the emission of a W^+ boson which decays into a neutrino$_\mu$ and an anti-muon.[40(p154)] Weak interactions involving the $W^{\pm}$ bosons do not preserve the electric charge of the particle containing the transformed quark. On the other hand, weak interactions involving the Z^0 boson do. The Z^0 boson interactions provide a way for neutrinos, which are only noticeably affected by the weak force and are not trapped inside atoms,[40(p7)] to interact with matter without swapping charges around.[40(pp114-115)]

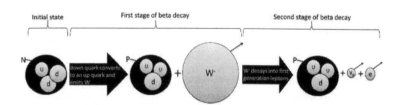

Figure 2

The $W^\pm$ and Z^0 bosons are massive force mediating particles. Because of this, the weak force is a short-range force similar to the strong force but different from the forces of electromagnetism and gravity which have infinite range. The $W^\pm$ and Z^0 bosons stand out because unlike the other fundamental force-transmitting particles, they have mass. Furthermore, the $W^\pm$ bosons stand out even more because they have charge while the other force particles do not. These unique features of the weak force-mediating particles indicated that there may be another particle that interacts with them to account for their high masses.[16,39(pp264-265),45] This is what led to the prediction and recent confirmation of the Higgs boson's existence. The Higgs boson cannot be detected directly but decays into high energy photons and the $W^\pm$ and Z^0 bosons, and it is the detection of these particles within particle colliders that indicate the Higgs boson's presence.[46] It is theorized that just like $W^\pm$ and Z^0 bosons, all matter particles interact with the Higgs field to some extent to acquire their mass. When these fundamental matter particles interact with each other via the fundamental forces to form larger systems of matter, other sources of mass appear such as the energy in the gluon field that keeps the quarks in clusters.[39(p262)] In general, however, the heavier the fundamental particle, the more it is believed to interact with the Higgs field. Therefore, each successive generation of fundamental matter particle interacts with the Higgs field more and more because as we descend through the generations, the masses of the particles increase. The small mass of the neutrino may be a sign that it does not interact much with the Higgs field.

Unification

Many physicists feel that there is a unity underlying the fundamental forces in the Standard Model. They theorize that at high-enough energy densities, like the conditions of the universe during the Big Bang, all of the Standard Model forces collapse into one; it's not until energy decreases that first the strong nuclear force splits off and becomes its own distinct force, then with a further decrease in energy, the weak nuclear force and the electromagnetic force split off from each other. In fact, well-established theory has been developed to demonstrate that the weak nuclear and electromagnetic forces can be unified within the same framework. Furthermore, this theoretical framework now has some observational support because it is dependent on the existence of the Higgs boson which has just been discovered at the LHC. Physicists hope to one day firmly establish the unification of this "electro-weak" force and the strong nuclear force. And of course, the holy grail of physics is the unification of all of the forces of nature—the three forces currently in the Standard Model and the force of gravity. This latter feat will most likely be accomplished by a "theory-of-everything".

Configurations are Information

A recent trend among scientists is to interpret the theories of physics within the context of information.[47] The key insight that enables this line of reasoning is that information is physical, i.e., ultimately, it is represented by the configurations of the fundamental particles that interact to form a single physical system.[48(pp2,56-57,87)] Information theory describes the communication that must take place in order for two distinct systems, and even the parts of those systems, to interact. Interacting parts and systems are all that exists in the universe so it may be the case that information theory plays a very important role in describing the nature of reality. As an interesting thought experiment, consider two systems in particular, one defined by humanity and the other by humanity's environment, which, ultimately, is at least the entire observable universe. Information theory should be capable of describing the communication, or, interaction between these two systems, interaction that enables us to construct theories that further our understanding of the universe. This type of reasoning has led some to believe that our theories in physics, such as thermodynamics, quantum mechanics, and relativity, are actually theories of information and therefore, they should govern how information behaves in the universe.[48(pp2,56-57,87)] Taking this approach has revealed much insight concerning the nature of reality and the remaining mysteries within physics.

In general, there are multiple configurations that a system can be in and still be considered as the same system (this is the concept of entropy). Each unique configuration provides the answer to the question: what is the state of each of the parts and the interactions taking place between them to form the system? In other words, each configuration corresponds to a certain amount of information. In addition to information, each configuration of a system also corresponds to a certain amount of energy, energy that is required to support the system's configuration.[37(p232),49(pp14-16,54-55)] All of a system's possible states can be organized according to energy, such as a list starting with the lowest possible energy state, the ground state, and ending with the highest possible energy states; each energy level will contain some number of the possible configurations of the system. If a system is to occupy a lower energy level, it will have to give up energy to its environment. On the other hand, if a system is to occupy a higher energy level, it will have to acquire energy from its environment. The transmission of this energy between the system and its environment must ultimately take place through one of the four fundamental forces. The more complex a system is, the more information it can represent, where in this sense, complexity is a function of the number of parts, the variety of parts, and the variety of interactions between the parts.[25(pp2-4),49(pp14-15)] Since energy is needed to support a system's configuration, the more complex the configuration, the more energy is needed to support that level of complexity. On a fundamental level, all matter is made of atoms, and within the atom, both the electrons and the nucleus can support a variety of configurations and hence, a variety of energy levels, complexity levels, and information content.

Information Processing

If a system's configuration represents information, then changes to the system's configuration represent information processing.[47] In nature, most of what we are directly aware of through human sensation is ultimately due to configuration changes of electrons within atoms. These particles can move up or down their energy levels—attain states of higher and lower energy, respectively—within an atom by exchanging electromagnetic radiation or heat with the environment. The minimum energy for an electron within an atom is known as the ground state, and in this state, the electron is securely trapped inside the atom. At the other extreme, if an electron is supplied enough energy, it can escape the attraction of the nucleus and become a free electron.[40(p37)] The activity of electrons in atoms is both a major source and sink of electromagnetic radiation because when electrons within atoms drop from one energy level to another, they release a photon whose energy is equal to the difference between the two energy levels. Likewise, if an electron absorbs a photon, it will rise to the energy level that accounts for the energy of the photon. The electromagnetic radiation resulting from electronic transitions within atoms include x-rays, ultraviolet rays, and visible light. Infrared waves result from the vibrational motion of atoms within molecules while longer and less energetic wavelengths, such as radio- and micro-waves, involve large motions of the charged particle, such as rotations or translations. The shortest and most energetic electromagnetic waves, those with gamma frequencies, are produced by the nucleus of atoms and high energy particle decays.

The nucleus of an atom is made up of protons and neutrons which are themselves made up of quarks. And similar to the electrons of an atom, protons and neutrons can become excited and enter higher states of energy due to the various spinning and orbiting motions of their constituent quarks.[40(pp82-83)] But unlike the electrons in an atom, the quarks will forever stay trapped within the protons and neutrons by the strong nuclear force.[40(pp82-85)] However, energy can be released by the nucleus in a variety of ways that are collectively known as radioactivity; this is essentially an atom's attempt to achieve a more stable state and it comes in the form of the nucleus releasing some combination of matter and electromagnetic radiation. For example, an alpha particle is the nucleus of a helium atom and it is emitted from the nuclei of larger atoms when they need to readjust their constituents to obtain a more stable state. Another form of radioactivity involves the highly energetic gamma rays that can also be emitted by the nucleus. Beta decay, on the other hand, is a form of radioactivity that is weak force-mediated and whose by-products include an electron and a neutrino$_e$ (first generation beta decay). The difference between beta decay and the other forms of radioactivity is that the by-products of beta decay, the electron and the neutrino$_e$, are not considered to have an existence within the nucleus.[40(pp37,40)]

The types of information transfer discussed above involve the electromagnetic and weak nuclear forces. Information is also transmitted via gravity and the strong nuclear force. Gravity is a large-scale force that becomes important for large agglomerations of matter; it can communicate information concerning the system's motion and mass. On the other hand, the strong nuclear force exists between quarks to ensure tight clustering; the intensity of gluon transmissions between quarks gets stronger the further apart they are. Both the weak and strong nuclear forces are short-range forces while the forces that are most apparent on the scales that we experience the universe are the electromagnetic and gravitational forces. On the larger scales of the universe, gravity and light reign supreme.

Atoms

Most of the mass of an atom is due to the nucleus which is usually a cluster of neutrons and protons, particles that do not possess a color charge and so therefore do not emit gluons to bind together. But recall that protons are positively charged. When these like charges attempt to come together to form the nucleus of an atom, they will have to somehow overcome the electromagnetic repulsion that they feel. It turns out that there is a residual strong force which is a by-product of the strong nuclear force affecting the quarks that cluster to form protons and neutrons.[40(pp7,40,105)] For the most stable atomic nuclei, this residual strong force is stronger than the electromagnetic repulsive force felt by the protons, allowing them to cluster together, and with neutrons, to form a nucleus. The residual interaction between nucleons is uncolored as opposed to the colored one that mediates the interactions between their constituent quarks, and it can be considered as the exchange of a virtual pion—a particle that is a cluster of one quark and an antiquark.[40(pp53,59)] The communications between the quarks forming a cluster through the gluon-mediated strong nuclear force, and the communications between the nucleons formed by these clusters through the residual strong force, occur in a very coordinated fashion so that the strong nuclear force continues to have an effect beyond the confines of the colored particles through uncolored interactions. Ultimately through this process, atomic nuclei are built.[40(pp40,105)]

The atoms known to exist can have anywhere from one to a little over 100 protons in the nucleus and tend to have the same number of electrons as protons, resulting in a physical system with a net electric charge of zero. Recall that for any physical system, there are many possible configurations that the system could be in and still be considered as the same system. This is true for the electrons in an atom because they can be configured in many different ways, each corresponding to a particular energy level. In general, Pauli's Principle is an organizing principle that governs how matter particles can come together to form a single physical system. It says that no two matter particles can occupy the same state within the same system.[40(pp91-92)] This principle governs the configuration of the electrons within atoms, and because of it, no two electrons can occupy the same state within the same atom. Stable states of the atom correspond to the states where electrons fill all the lowest available energy levels in such a way that they obey Pauli's Principle. The outer electrons are available for bonding, or, being shared by other atoms. Furthermore, electrons play an important role in the interactions of the atom with its environment, and through the Pauli Principle, electrons give objects their form and bulk.[50] The size of an atom is determined by the size of the nucleus and the energy states that the electrons occupy—its size grows as its electrons occupy higher energy levels. If an electron is supplied enough energy, it can escape the attraction of the nucleus altogether and become a free electron.

The atomic elements are organized according to their weights in the periodic table of elements. When they are listed from the lightest to the heaviest, elements with similar properties recur at regular intervals and turn out to be groups that have the same number of electrons in their highest occupied energy level.[40(pp13-16)] All "normal" matter in the universe, from molecules to macroscopic objects, are ultimately built from atoms that have bonded together by sharing electrons.

The atoms of all of the elements can have varying amounts of neutrons in the nucleus. One of the functions of the neutrons is to stabilize the nucleus, something it does in at least two ways: 1) through its influence on the electromagnetic repulsion experienced by the protons in the nucleus and 2) through the electroweak force experienced throughout the atom. As the number of protons increases, the combined electromagnetic repulsion will soon become stronger than the binding ability of the residual strong nuclear force. The number of neutrons that exist within the nucleus can exceed the number of protons because the presence of the neutrons acts as a buffer to the repulsive electromagnetic force acting between the protons. This helps to create a stable nucleus but the larger the difference between the numbers of neutrons and protons in the nucleus, the more unstable the nucleus becomes. As the number of protons increases, their number eventually becomes too large to support a stable nucleus. In these cases, the nucleus achieves a more stable state through radioactive means. This process results in one element transforming into another by emitting or absorbing particles, or by protons and neutrons turning into each other, as in beta decay. This type of condition is typical of atoms starting at uranium in the periodic table of elements.[40(pp34-37,42)]

Communication is Key

In this chapter, we looked at the most fundamental pieces of matter and energy in the universe, the fundamental matter and force particles in the Standard Model. We saw that the interactions of these particles are ultimately responsible for all the matter and complexity that we observe in the universe. And it turns out that information is a concept that can be used to describe nature since information essentially is represented by the configurations of the fundamental particles in matter and the communications that take place between systems of particles.

This chapter laid the foundation for your ability to recognize correspondences that exist between the particle relationships in the Standard Model of Particle Physics and the circuitry within key areas of the brain.

We will now turn our attention to quantum mechanics, the theory out of which the Standard Model is rooted. When considering quantum mechanics, we encounter situations in which space and time can no longer be considered as being continuous, and physical systems become completely isolated from their environment. When this happens, the system is no longer communicating with its environment and begins to exhibit counterintuitive characteristics.

Chapter 3

Quantum Mechanics

A Different World

The nature of reality on the lowest levels of the spatiotemporal hierarchy—the ones that are populated by the particles in the Standard Model—is much different from our world, the one we experience on a daily basis.[48(p195)] In our world, space and time are seen as being continuous because the states of systems change smoothly and don't jump discretely from one to the other like a picture in a flip book. In this world, particles appear to be point-like chunks of matter as they move smoothly through space and interact with other objects, and in order for any message to be sent from one location to the next, matter will have to be transferred between the two locations. This message can be comprised of either matter or force particles, and since the speed of light is the fastest anything can travel through space, it is the fastest any kind of communication can occur in the universe.

But through a thoroughly validated theory called quantum mechanics, which the Standard Model is rooted in,[21(p80),39(p352)] we know that the way we perceive physical reality to be is, in general, not the way it actually is on the smallest levels of the spatiotemporal hierarchy; on these levels, the states of the universe and the systems that comprise them are actually discrete. This represents a fundamental difference between how things are on the lowest levels of the spatiotemporal hierarchy and how they are on the levels that we are most familiar with. Two additional differences are that all matter has wave characteristics and that the universe is actually nonlocal, meaning that an action in one location can have an effect on another before light has even had time to traverse the distance separating them. It turns out, though, that as we ascend the spatiotemporal hierarchy, the quantum mechanical descriptions of the systems in question actually reproduce all of the common sense views of the nature of reality listed above. That is, the quantum laws join seamlessly with the classical Newtonian laws (those of Sir Isaac Newton) describing systems on the scales of the spatiotemporal hierarchy that we are most conscious of.[21(p199),39(p186)] The principles of quantum mechanics — the way things are on the smallest scales of the universe — seem so alien to us because they do not manifest in the macroscopic objects that we perceive in our everyday, or, classical world.[48(p195)] And just as in the previous chapter where information helped us characterize a system's configurations and dynamics as well as the communication that takes place between systems, information is the crucial concept that connects what we know about the smallest scales of the universe and what we know about the macroscopic ones.[48(pp178-179,195),51]

Discrete States

All physical states of the universe are discrete, and when considering them, it can be useful to distinguish between empty space and the matter that exists within it, because in modern physics, even empty space itself is a subject of inquiry, one that is carried out both experimentally and theoretically. Our theories allow us to probe scales of empty space smaller than our experimental techniques currently can and they tell us that on the smallest spatial scales, spacetime itself is actually discrete, meaning, there is a limit to the number of times that you can chop it up because eventually, you'd reach a minimum distance, known as the Planck length, that defines the smallest chunks of spacetime and determines the amount of information that can be put into a region of space.[47] But this doesn't necessarily mean that distances shorter than the Planck length do not exist, only that it is impossible for us to distinguish between two locations separated by a distance shorter than the Planck length. It is possible that there are dimensions of spacetime smaller than the Planck length which means that at every discrete chunk of spacetime, there may be additional independent directions in which the fundamental particles can travel, but we are theoretically prohibited from observing their motion due to the constraints of the Planck length.

Spacetime itself isn't the only thing that is discrete since the internal configurations of systems are as well—fundamentally, the particles comprising these systems have properties that change discontinuously. One of these properties is their intrinsic angular momentum states known as spin. Another is their energy, which changes by discrete amounts called quanta that typically correspond to the fundamental force particles. In fact, quantum field theory says that each particle in the Standard Model belongs to its own universe-filling field. In this view, when a particle, say, an electron, exists in the universe, it is an excitation of the underlying electron field.[16,21(pp74-76),47] This is a concept that joins nicely with the theoretical notion that spacetime itself is composed of discrete chunks, like the pixels of a digital image. All of these discrete chunks represent locations in the universe where the fundamental particles can become manifest and physical processes can occur. If a fundamental particle exists anywhere in the universe, then the field for that particle is excited at that particular chunk of spacetime. Thus, the fields of the fundamental particles in the Standard Model are also discrete. In quantum theory, each particle of a particular type, or, each quanta of a particular field, is identical and the only thing that distinguishes them is their state, defined by such things as their location, energy, and spin.[48(p224)]

Recall that there is a fundamental difference between matter and force particles regarding the number of each particle type that can simultaneously exist in the same exact state—each matter particle must occupy a unique state while there is no limit to the number of a particular force particle that can occupy any given state. For example, when multiple matter particles of a particular type occupy the same physical location, such as the two up-quarks in a proton, there must be some other property that can be used to distinguish the particles. For the case of quarks inside nucleons, this property is color. On the other hand, for two electrons that exist together in the lowest possible electron energy configuration in an atom, the spin of the electrons can be used to distinguish them.

It is the discrete nature of the universe that allows information theory to be applied to quantum systems because their unique configurations represent information. As a system evolves over time, the micro state changes of any component particle of the system, such as flips in a particle's spin state, represent bit flips.[47] Systems communicate these changes via the four fundamental forces, although gravity does not seem to play a role on the tiny scales of most quantum systems;[40(p47)] these changes result in the system emitting fundamental force particles into its environment that convey information about its state changes, and all systems that interact with these force particles have access to this information. This may not seem like a big deal when only considering inanimate objects in the environment of a force-radiating system, but as we will see shortly, it is a very important aspect of nature. In addition, through some form of physical sensation, all life forms are sensitive to some of these force signals, allowing them to interact with, and evolve as a consequence of, their environment. In fact, humans are able to extract higher-order information from these signals to develop theories about the nature of reality, i.e., humans can use these signals to form mathematical and conceptual models of the universe.

Waves of Probability

The wave properties of matter become apparent when a physical system is sufficiently isolated from its environment, a condition that exists when communications via the fundamental forces cease between it and its environment.[39(pp210,212)] It is usually the most simple, least energetic, and tiniest systems that display quantum characteristics:[48(p210),52] their low energy and simple configurations limit internal changes and result in reduced emissions of fundamental force particles into the environment, and their small size limits the amount of interaction that they have with environmentally-borne fundamental force particles.[48(pp208-209)] For these reasons, large periods of time go by in which these types of systems are in complete isolation. There is no specific size that a system must be to display quantum characteristics, and within the laboratory, scientists have been successful at getting larger and larger objects into quantum states due to their improved ability to sufficiently isolate them.[48(p173)]

When this happens, we can no longer view the system as a physical entity having any one particular configuration, but rather as being represented by a nonphysical, ethereal probability wave that expresses the likelihood of possible system configurations. This probability wave embodies all that we know about the isolated system so that the best we can obtain is a probabilistic description of it and cannot say with any certainty that it will be in any one particular configuration when measured.[21(p197),39(p91)] In contrast to the classical view where a particle is a point-like chunk of matter with a well-defined position, the quantum probability wave associated with a particle is spread throughout the entire universe, usually having a magnitude approximately equal to zero outside a very small region. It may help to view this probability wave as existing within the underlying fields of the fundamental particles in the Standard Model. Therefore, when all of the implications of quantum mechanics are considered, we see that fundamentally, each species of matter and force particle in the Standard Model are actually modeled as a combination of point particle and probability wave that both manifest in the underlying space-filling field for that particular particle species;[21(pp75-76,80),39(p256)] the form the fundamental particle takes — particle or wave — depends on whether or not information about it is being gathered by its environment. The same applies for all larger physical systems formed from the interactions of the fundamental particles in the Standard Model.

The probability wave is also referred to as the wave function of a system and is completely theoretical since it is believed to be impossible to directly measure and observe. Therefore, unlike the mathematics of classical physics, not all of the mathematical quantities in quantum mechanics correspond to something that we can directly observe in the physical world.[21(pp211-212)] Quantum mechanics started as a theory rooted in the real world but it then grew to incorporate nonphysical entities, i.e., entities that model aspects of the universe other than physical objects. These nonphysical aspects, namely the probability wave, are closely related to the physical ones that we can observe.[39(pp89-91),53] This has allowed for our continued use of these purely theoretical-seeming concepts to describe quantum systems, a method that has been verified through countless experimental observations in the years since the development of quantum mechanics in the first quarter of the 20th century. For many, whether or not this wave is a real physical manifestation existing somewhere in the universe, and not just a mathematical construct, is still an unresolved issue to this day.[21(pp169-170,236),39(pp89-91),48(p255)]

In quantum theory, the probability wave for a system can travel through the environment and interact with objects, making it possible for it to be inferred through the identification of wave-like phenomena such as interference and diffraction. These wave-like phenomena build up whenever the environment makes repeated measurements of an ensemble of systems, causing their description to instantly go from the probability wave to a classical view where they definitely exist in a single configuration, an event referred to as the collapse of the wave function.[21(pp200-201),39(pp87,183-184),52,54] Over time, these collapses can reveal interference effects that may be interpreted as the many different possible histories of the system leading up to the measurement all playing an important role in the measurement outcome. In other words, the wave phenomena associated with a quantum system are manifestations of the interaction of all possible system configurations prior to measurement.[39(pp183,212)] Recall that as a system becomes more and more classical, the quantum description joins seamlessly with the classical Newtonian description of the system, an indication that the latter is an approximation of the former for macroscopic systems. This makes interference patterns a hallmark of quantum systems since they do not occur for classical ones. However, classical systems may not display interference phenomena but they do have wave functions nonetheless. As opposed to quantum systems whose wave functions are relatively broad compared to the size of the system, classical systems have wave functions with a very narrow peak, much smaller than the system size, that coincides with the trajectory predicted by classical physics.[21(p199),39(p183)]

The Uncertainty Principle

As we saw above, physical objects are not the only things that display quantum characteristics on the smallest scales of the universe; even empty space itself behaves counterintuitively on these lowest levels of the spatiotemporal hierarchy. In addition to spacetime ultimately being discrete, matter can temporarily appear for short periods of time in otherwise completely empty space, a phenomenon that boils down to a lack of measurement like the quantum characteristics displayed by completely isolated physical systems.[39(p306)] This attribute of empty space on the smallest scales of the universe is due to the Uncertainty Principle which says that there are pairs of observables that cannot be simultaneously measured to arbitrary precision, such as position and momentum, or, energy and time. For example, if the position of a quantum particle is measured precisely in a given instant of time, its momentum in that instant—the speed and direction of its motion—will become infinitely uncertain. The reverse holds as well—if you precisely measure the momentum of a quantum system, the result is complete uncertainty in the system's position. Essentially, the Uncertainty Principle says that this is a fundamental aspect of reality since it holds regardless of whether any measurements of the quantum system are being carried out or not[48(p204)]—ultimately it has to do with the wave nature of quantum mechanics.[39(p98)]

Brian Greene—theoretical physicist well known for his work popularizing string theory—provides an analogy that illustrates the irreducible amount of uncertainty in quantum measurements. He likens the uncertainty in the exact position and momentum of a particle at any given time to the amount of uncertainty in the same quantities for an object imaged by a camera that takes a series of snapshots. If the shutter speed on the camera is slow, there will be a high degree of uncertainty regarding the object's position but its momentum will be well known. If the shutter speed is fast, the object's position will be well known but there will be uncertainty in its momentum.[21(p31)] In quantum mechanics, the relationship between the position and momentum information is the Fourier Transform.[54]

Earlier, when discussing forces felt by matter, we encountered an example of energy-time uncertainty in the short-range nature of two of the nuclear forces. This uncertainty relation says that the larger and more energetic the matter that forms out of the vacuum of space, the shorter the amount of time it exists. Therefore, since the virtual pion that is shared between nucleons is a massive particle, it is short-lived, resulting in the short-range nature of the residual strong nuclear force. Similar reasoning involving $W^\pm$ and Z^0 bosons reveals the short range nature of the weak nuclear force as well.[40(pp26,54)]

Nature's Measurements

There are fluctuations in the ambient energy of empty space that result in the temporary manifestation of pairs of virtual particles whose collective properties obey conservation laws such as angular momentum and energy. These experimentally-confirmed virtual particles arise everywhere within the universe and play an important role in the measurement of quantum systems because when they arise, they can interact with quantum systems and disseminate information regarding them into the environment before disappearing again into the vacuum of space. This causes the wave function of the measured quantum system to collapse, allowing for the manifestation of one of the possible states that it could have been found to be in.

These virtual particles are the universe's way of measuring itself, even in the deepest vacuum and even when the quantum system is in its lowest energy configuration so that it is not radiating light.[48(pp203,205-206)] Therefore, even if we configure a tiny, simple system in its lowest energy state and isolate it from the environment, which includes everything from the detectors of scientific equipment to atoms and molecules floating through the air, it can still be measured by the universe through virtual particles that continuously and ubiquitously pop into and out of existence. Consequently, it must be kept in mind that the presence of humans and their equipment are not necessary for a measurement to be carried out on a quantum system—the physical environment is all that is required.[39(Ch7),48(Ch7)]

So with all things considered, we have that an isolated quantum system can interact with, and hence, be measured by, objects within its environment that exist on any level of the spatiotemporal hierarchy—individual fundamental particles, atoms, detectors, etc. Before measurement, all possible system configurations are encoded in the probability wave and interact with each other and the environment through wave mechanics; after measurement, the probability wave collapses and only one of these possibilities manifests while the others cease to exist and no longer interact with the system which is now in physical form. The possibilities that cease to exist have essentially been filtered out by the measurement of the quantum system. In this sense, the probability wave is like a signal consisting of possible system configurations multiplexed together, and the measurement of the signal filters it so that only one possibility manifests. For classical, macroscopic objects, there is one configuration that is much stronger than the other possibilities — the one corresponding to the classical Newtonian state of the system. For quantum systems, however, there are multiple possible configurations that each have a significant chance of being the outcome of a measurement. It's like for small, simple, and low energy systems, none of the signals representing the possible outcomes of the experiment have high enough strength to ensure that it will be filtered out of the superposition of possibilities by a measurement, a situation analogous to the detection of low power signals in noise using a filter-based detection technique,[55,56] and in fact, the mathematics describing the two processes is very similar.[57,58] Furthermore, when the mathematics describing quantum systems and signal processing is compared, it becomes apparent that the Uncertainty Principle for complementary variables in quantum mechanics — the types of variables that cannot be measured simultaneously to arbitrary precision — corresponds to the inability to multiplex signals in the theory of signal detection.

Example: Double-slit Experiment

The double-slit experiment can be used to introduce the most essential ideas of quantum mechanics, and in it, beams of particles are made to display the wave characteristics of matter (Figure 3). Recall that while not considered to be a real physical entity, the probability wave of an isolated quantum system, such as an electron, can interact with matter and be made to interact with itself through interference, a phenomenon that can be viewed as the possible configurations of the electron leading up to the measurement—its possible histories—interacting with each other. Interference occurs when probability waves with the same, or nearly the same, frequency occupy the same region of space and combine to form a resultant wave with greater or lower amplitude in some places than the original waves. This is a phenomenon that is never observed in classical objects because at any moment in time, they occupy one, and only one, configuration.

The first stage of the electron double-slit experiment is the production of a beam of electrons that are isolated from any type of measurement and that travels towards a screen having two small, closely spaced slits in it. The second stage of the experiment is the propagation of the electrons that have passed through the slits to a screen used to detect them. In the classical view, each electron detected at the detector screen could have only traveled through one of the slits, yet when the experiment is run again and again, one electron at a time, an interference pattern builds up at the detector screen indicating that wave phenomena describe the underlying processes of the experiment (Figure 3a). Classically, we would expect many hits in only two locations: behind slit one and behind slit two. Quantum theory accurately describes the experiment because it treats the unmeasured electron as a probability wave, one that can interact with the experimental apparatus and also interfere with itself. This probability wave describes an ensemble of identically-prepared electrons being sent through the double-slit apparatus so it is not entirely accurate to view each electron as if it has its own unique probability wave, rather, each electron in the double-slit experiment is described by the same probability wave (Figure 3b). When only one electron is sent through the system, only one location on the detector screen will register a hit, signaling the location of the electron, but when the experiment is run over and over again, the interference pattern will eventually build up at the detector screen. Every single electron that passes through the system will arrive at a location on the detector screen that is in accord with the interference pattern predicted by quantum mechanics, but will only strike the screen in one location; it requires running the experiment many times to observe the interference pattern[39(p87)] (Figure 3c).

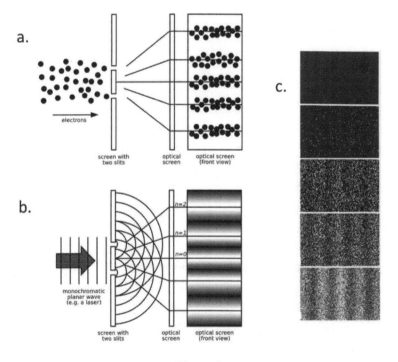

Figure 3

The two slits in the first screen provide two possible paths, or histories, that any one electron can take on its way to the detector screen. But when the electrons are sufficiently isolated from their environment, a probability wave describes their interactions with the experimental setup. The two possible paths cause interference in the probability wave that manifests at the detector screen when the experiment is repeated many times with identical electrons. Both possibilities—each of the possible histories—contribute to what we observe.[39(p179)] If we pick one of the slits to measure the passage of electrons through, we destroy the interference because we will have eliminated one of the possible slits that the electron could have traveled through by forcing the wave function to collapse. For interference to occur, the electrons must remain isolated until measured by the detector screen, underscoring the important role the transfer of information from a quantum system to its environment plays in the manifestation of the classical world in which we live.[21(pp200-201),39(pp210,212)]

Nonlocal Universe

In the everyday world that we perceive, the classical world, if we are to exert an influence at a location other than the one we are at, we have to physically send a message from our location to the other, and this communication cannot occur instantly since the speed of light is the fastest that anything can travel in the universe.[39(pp12,79-80,199)] For quantum systems, the situation can be very different. When two distinct quantum systems have interacted at any time in the past to form a single quantum system, correlations in their possible configurations can result, a phenomenon known as entanglement. In this situation, the two quantum systems become information-theoretically linked, meaning that when it comes to measurement, the two distinct quantum systems are treated as one, no matter how far apart they are physically separated.[39(pp114,122),48(pp175-177)] Measurement of one system simultaneously collapses its wave function and the wave function of the other system, and both occur in such a way that the properties of the systems obey conservation laws, such as angular momentum.

Entanglement makes it appear as if information were

instantly transmitted from one system to the other so that the receiver would know how its wave function should collapse to ensure that its properties are properly correlated to the sender's. But it is really an indication that the prior interactions of the two quantum systems make them, in the eyes of the universe, one single quantum system, even though they are physically separated—the probability waves experience a correlated probabilistic collapse, no matter which system actually experiences the measurement.

An example of entangled quantum systems is the vacuum fluctuations that arise due to energy-time uncertainty. If one of the particles is measured and found to be spinning clockwise, then we know with certainty that the other particle is spinning counterclockwise—no measurement need be made. Since angular momentum must be conserved between the two virtual pairs, we instantly know what the properties of both particles are after a measurement of just one of them. Even though the collapse of the wave function for any one of the virtual particles is probabilistic and doesn't occur until it is measured, the other particle's wave function still manages to instantly collapse in such a way that the particles' properties are correlated, indicating that on a fundamental level, the universe is capable of displaying nonlocal phenomena.

Extreme Communication

The systems that exist in the universe have properties that vary widely in terms of size, dynamics, and interactions with the environment. In this chapter, we examined extremely tiny systems, ones that are low energy and interact little with their environment via the four fundamental forces. On these small scales, we saw that both spacetime and the configurations of systems are discrete. We also saw that all matter is associated with a nonphysical probability wave that determines the likelihood of the possible configurations that may result from a measurement of the quantum system. In addition, we saw that humans and their scientific equipment are not required for quantum systems to be measured; any physical object—from virtual particles produced by quantum fluctuations to atoms floating through the air—can interact with a quantum system and force its probability wave to collapse and manifest one of its possible configurations.

This chapter prepares you to recognize the parallels that exist between the quantum nature of the universe, and the ways in which our brains appear to construct the reality that we perceive.

It turns out that systems which are tiny, low energy, and isolated are not the only ones that are capable of displaying counterintuitive characteristics. By studying intense gravitational fields and extremely fast moving objects, we have uncovered additional insights regarding the nature of physical reality. In the 'next chapter, we will explore some of these insights as we consider relativity theory.

Chapter 4

Relativity Theory

Classical Madness

In the last chapter, we saw that the classical Newtonian description of our everyday world is a macroscopic approximation of quantum mechanics—the laws that describe very tiny, low energy, and isolated systems. It turns out that the Newtonian description is also a low speed and low mass approximation of the laws that govern extremely fast moving or extremely massive systems[39(p10)]—the laws of relativity theory (Einstein's special and general theories of relativity). Just as we find counterintuitive phenomena when observing systems existing on the lowest levels of the spatiotemporal hierarchy, we again find counterintuitive phenomena when observing systems traveling at speeds close to the speed of light and also when observing systems involving massive objects that produce extremely strong gravitational fields. However, unlike quantum mechanics, relativity theory does not incorporate probabilistic concepts and is therefore a classical theory like the Newtonian description of our everyday world[39(p329)]—it deals with systems that do not become completely isolated from their environment. The counterintuitive phenomena in relativity theory have to do with the nature of space, time, and gravity. And just as information is the crucial concept that connects our view of the smallest scales of the universe to the intermediate ones where our everyday world resides,[48(pp178-179,195-196)] we are able to use information theory to draw conclusions from the findings of relativity theory and arrive at insight regarding the nature of reality on the larger and more energetic levels of the spatiotemporal hierarchy.

Space and Time are Relative

Recall that ultimately, all physical systems are a result of interactions between the fundamental matter particles via the four fundamental forces and that the possible sizes of these systems range from the size of elementary particles in the Standard Model to the size of the entire universe itself (Figure 1). Furthermore, systems existing on one level of the hierarchy result from the interactions of systems existing on levels below, and all systems, except for perhaps the universe itself, have an environment that they are a part of and can interact with. At any instant of time, these systems have both an internal and an external configuration. The internal configuration is determined by the microstates of the component parts of the system and the external configuration is the relation of the system as a whole to its environment. Examples of things that define a system's internal configuration include spin orientations of the constituent particles, color charges of the quarks, energy levels of electrons, etc. A system's external configuration is determined by its motion through its three-dimensional environment, something that ultimately results from its interactions with objects in the environment, such as electromagnetic or gravitational influences.

Special relativity says that every physical system has its own internal clock. Since time is seen through change,[39(pp141,225-226)] it follows that the rate at which time elapses in a system's internal clock reflects the rate that its internal configurations change. Special relativity also says that the rate of change of a system's internal and external configurations—how fast its clock ticks and how fast it moves through space, respectively— are complementary.[39(pp48-49)] This means that the faster a system moves through space, the slower its clock ticks and vice versa. In addition, one of the key insights of special relativity is that the speed of light is constant and is the speed limit of the universe. Therefore, the speed of light is the fastest that information can travel[48(pp120,142)] which is important because all interactions between systems are ultimately some form of communication, i.e., information transfer.

We perceive the universe to be four dimensional—three dimensions of space and one of time. Normally, we consider a system's motion to take place within the three spatial dimensions but it can also be considered to travel through time as well. Special relativity says that the combined speed of a system through space and through time is always equal to the speed of light.[39(pp48-49)] Consequently, a truly stationary system does not travel through space but it does travel through time and does so at the speed of light. The significance of this is that when a system is stationary, changes to its internal configurations occur at the fastest rate, i.e., the system will age fastest. When the system begins to move through space, its internal changes will occur slower—the faster the motion, the slower the system changes internally, and hence, the slower it ages and travels through time. At the opposite extreme, time stops for a system that is traveling through space at the speed of light because all of its light speed travel through time has now been converted into light speed travel through space.[39(p49)] For such a system, internal changes no longer occur. In reality, only massless particles, like photons, can travel at the speed of light since physical systems with mass can approach the speed of light but never actually reach it—the faster they travel, the larger their mass becomes, making it harder and harder for them to travel at light speed.

One implication of the speed limit of the universe and the internal clocks of all physical systems is that space and time are actually aspects of a unified concept called spacetime. For two observers whose states of motion differ, their perceptions of spacetime will also differ from each other but will adjust in such a way that both observers agree on the speed of light.[39(p47)] And to ensure that this speed stays constant and is perceived as being the same by any observer, no matter the details of their motion, distances must change to adjust for motion-induced changes in their clocks. Therefore, two observers whose states of motion differ can perceive the same event in two different, yet, equally valid ways since an event only occurs from their perspective when information about that event gets to them.[48(pp134-135,140)] Note that although perceptions of spacetime are relative and depend on an observer's motion, relativistic effects can never reverse causality, meaning, two observers undergoing different motion will perceive events occurring at different times, but they will not see consecutive events occur in a different order.[48(p141)] Furthermore, relativistic effects only become significant when a system travels through space at speeds close to the speed of light; therefore, these effects are foreign to us because the systems existing in our everyday world have relatively slow speeds.

Gravity's Influence

Another implication of the universe having a speed limit and physical systems having internal clocks is that both stationary systems and those moving through space at a constant speed have trajectories in spacetime that are straight, while the trajectories of accelerating systems are curved. It is not hard to see this when we observe objects that move through space in a curved manner, but it also applies for objects that we observe to be accelerating along a straight path through space—in this case, the curve is not in the three spatial dimensions but occurs along the dimension of time. The Equivalence Principle in Einstein's general theory of relativity says that the effect of gravity on a physical system is the same as that of acceleration—it causes the trajectories of systems through spacetime to curve.[16,39(pp68-69),47] In this view, the fabric of spacetime itself is warped and curved by the force of gravity.

Recall that gravity is produced by mass and is an attractive force that acts on all matter and light. Although not yet confirmed, the Standard Model predicts that this force is mediated by a massless particle called a graviton that travels at the speed of light and gives matter a gravitational field that spans the universe. If this turns out to be correct, then these particles somehow interact with objects in the environment of their source and cause them to accelerate, or in other words, to travel a curved path through spacetime. Recently, instead of being viewed simply as the presence of mass causing spacetime to curve, gravity, and hence, the would-be gravitons emitted by massive objects, is increasingly being viewed as emerging from physical processes of the constituent particles of matter, analogous to basic thermodynamic processes.[59]

Gravity is the one force that we know of that is not yet officially included in the Standard Model and there are some key differences between it and the three forces that the model does account for. The strong nuclear, weak nuclear, and electromagnetic forces act in a discriminatory manner since some types of particles feel them while others do not. In fact, even two different types of particles that feel the same force may feel it to varying degrees. Gravity is different because it acts attractively on all objects in the universe.[40(p7)] Furthermore, you cannot shield yourself from it like you can the other forces,[39(p65)] and on small scales, where the other three forces are important, gravity does not play a significant role. On the other hand, in the case of bulk matter, the short-range nuclear forces are no longer important and the net electric charge of objects is usually zero. Therefore, gravity takes over as the dominant long-range force, especially for large objects such as planets, stars, galaxies, etc.[40(pp6,47)]

As stated earlier, gravity is increasingly being viewed as information related to physical processes occurring within matter. This information gets communicated to objects in the environment and the interaction manifests as organization.[37(p232)]

The organization that results from gravity's influence is the increased entropy of the environment, i.e., gravity causes objects to take up more probable configurations by traveling along curved trajectories through spacetime.[59] This type of organization represents increases in entropy whenever gravity is significant. On the other hand, when gravity can be neglected, increases in entropy appear more like the manifestation of disorder within systems, such as the random spreading of atoms of a gas that were initially bunched together inside a container, or the disintegration of matter over long periods of time.

Invisible Galactus

All objects with mass produce gravitational fields that curve spacetime and slow the internal clocks of objects in their environment when they accelerate them in ways that increase the environment's entropy.[21(p14),39(p173)] Black holes are the most massive objects in the universe; therefore, they produce the most powerful gravitational fields. For one to form requires high energy processes, such as the collapse of an extremely massive object due to the force of its own gravity, or perhaps the collisions inside the most powerful particle colliders — in either case, the result is an infinitely small point called a singularity. Since black holes do not emit electromagnetic radiation or matter in the classical sense and are essentially just pure gravity, they are practically featureless.[59] And consistent with gravity increasing the entropy of the environment, black holes also happen to be the objects in the universe with the most entropy,[39(p173)] entropy that is a consequence of the lack of observable attributes other than their size and angular momentum, both of which are inferred from their gravitational influence on objects in their surroundings.

Gravity is an attractive force, and since black holes produce the strongest gravitational fields, they are really good at attracting other objects. Once an object passes through the event horizon, an immaterial spherical boundary surrounding the singularity that is considered the edge of a black hole, it is lost to the rest of the universe forever—just as a black hole does not emit any light or matter because of the strength of its gravitational field, nothing that falls into one, whether it be light or matter, can ever get out.[39(p173),59] When an object approaches the event horizon, it is accelerated closer and closer to the speed of light by the intense gravitational field of the black hole and we never actually see the object fall in because its clock will move slower and slower and the wavelength of the light that we use to see it will become longer and longer as it fights to escape the gravitational pull until eventually, the object freezes at the event horizon at the same moment that it stops emitting electromagnetic radiation[60,61]—time comes to a halt at the edge of a black hole.

Once an object falls in, its mass and angular momentum is transferred to the black hole. The mass of the object affects the black hole in two ways: it goes into increasing both the black hole's gravitational field and the surface area of its event horizon. Because black holes are so featureless, once one has formed, there is no way to tell what fell into it; i.e., all black holes the same size, no matter how they formed, will be practically identical to each other.[47,48(pp231-232,238)] But recall that all physical systems also represent information. The rest of the universe loses the information embodied by an object that falls into a black hole, but the information is not totally lost since it is captured by the black hole's event horizon whose surface area increases by an amount that equals the information content of the object that fell in—information resides on the black hole's surface, its event horizon.[61] The surface area of the event horizon corresponds to the black hole's entropy—the number of states that it could be in and appear the way that it does. The information represented by the captured object adds to the number of possible configurations the black hole could have but that we know nothing about because it is so featureless.[21(pp253-254),39(p173)]

In the classical description of a black hole, matter and light can be received but none emitted. However, the quantum view makes it possible for particles to be emitted due to Hawking radiation, a phenomenon resulting from pairs of virtual particles—a particle and its antiparticle—produced by vacuum fluctuations at the edge of black holes.[47,48(p233),59,61] The antiparticle falls in while its partner falls out. When separated like this, these particles become real; the negative energy antiparticle contributes to the evaporation of the mass of the black hole[21(p249),48(pp233-234),60] while the partner is radiated away and loses energy fighting against the gravitational pull of the black hole. In short, Hawking radiation is a quantum phenomenon that is responsible for the evaporation of black holes.

Black Holes and Holography

When we consider a black hole's role in the universe as an information processor, we learn that the universe may in some ways be analogous to a hologram, which, in the typical sense, is a two-dimensional surface on which the information necessary to optically construct a three-dimensional image of an object is stored nonlocally, i.e., all of the object's 3D information is stored everywhere on the 2D surface of the hologram.[39(p482),47,62] To see this, reconsider for a moment just what a black hole is: when more and more matter is packed into the same region of space, the matter will eventually collapse in on itself thanks to its own gravity, creating a black hole which can be viewed as a singularity that produces an intense gravitational field within which exists an immaterial 2D spherical surface marking the boundary beyond which no light or matter can cross and continue to send information other than gravitational influences to an observer positioned outside the black hole. This boundary is the event horizon and is regarded as the location where the configurational information of the matter that falls into the black hole resides — the surface area of a black hole is the entropy of the black hole. So from our perspective, the configurational information of the largest amount of matter that could possibly fit into any volume within our three-dimensional universe resides on a two-dimensional surface bounding that volume. This is the Holographic Principle[21(p255),39(pp481-482),47,63] and note that when a black hole is examined from a computational standpoint, we see that all of its component parts function coherently as a single entity and information is stored nonlocally within it,[47] observations that are completely consistent with holography in general, as is the recent finding that the event horizon can be viewed as a fractal.[64] This Holographic Principle may also apply to the whole universe so that it too may have a boundary somewhere that contains all of the information underlying our reality.[39(p482),47,62]

There's Matter Veiled

We have just seen that counterintuitive phenomena not only exist on the smallest scales of the spatiotemporal hierarchy where tiny, low energy, and isolated systems tend to exist, but they can also become manifest on larger scales when physical systems either travel at speeds close to the speed of light or are extremely massive and produce extremely intense gravitational fields. Systems occupying these types of states are described by Einstein's relativity theories — special and general — whose key insights reveal much about the nature of reality.

Special relativity says that each physical system has an internal clock whose rate of elapsed time depends on how fast the system travels through space. Furthermore, because the speed of light is the constant speed limit of the universe, no matter the motion of the observer, lengths are also relative and adjust to compensate for the motion-induced changes to the observer's internal clock. In addition, we saw that gravity is described by general relativity and may also be a product of physical processes occurring within matter.

This chapter prepares you to see similarity between what Einstein's relativity theories tell us about the universe and the full implications of our brains consisting of several neuronal oscillators whose coherence and rate of oscillation can vary across a wide range of timescales.

Next, we will examine the large-scale structure and dynamics of the universe and will discover that it consists of much more than the matter we see with our eyes or with our telescopes. There is mysterious dark matter and energy not currently in the Standard Model that influences the normal matter of the universe, matter produced by the interactions of the fundamental particles that are currently in the Standard Model.

Chapter 5

Large-scale Structure and Dynamics of the Universe

The Cosmic Web of Matter

When we began our discussion about the nature of physical reality, we considered the lowest levels of the spatiotemporal hierarchy—the scales at which fundamental particles interact and form larger particles and atoms. We saw that if these systems become isolated from their environment, they no longer exist physically but continue to interact with physical systems in the form of immaterial probability waves that determine the likelihood of them manifesting a particular configuration when measured. Then we considered classical systems of all sizes—from the small fundamental particles to large macroscopic objects—that are moving extremely fast, and ones that produce extremely intense gravitational fields. We saw that black holes are the most extreme of such systems because their presence warps spacetime to such an extent that they bring time to a complete standstill, and they create the highest entropy systems in the universe because they themselves are the objects with the highest entropy and they increase the entropy of their environment through the effect of their intense gravitational fields. We will start our discussion of the large-scale structure and dynamics of the universe by considering a type of physical system that tends to have black holes as their heart and soul—galaxies.[65]

Super massive black holes are believed to reside at the centers of most galaxies and produce far-reaching gravitational fields that cause matter—stars, planets, moons, dust clouds, gas clouds, etc.—to orbit around them. Individual galaxies also interact with each other gravitationally to create larger structures such as galaxy clusters, linear arrays of galaxies called filaments, and two-dimensional arrays of galaxies called walls. These systems of galaxies interconnect to form a cosmic network, or, "web" of matter in the universe. In addition, very large empty regions called voids form between these structures.[66,67]

Currently, there is some debate about the distribution of matter in the universe on the largest spatial scales. On one hand, we have our conventional model of the universe that is based on Einstein's equations; this model assumes that on the largest spatial scales, the distribution of matter is smooth. In this view, the matter in the universe is clumpy only up to a certain point, such as the systems of galaxies listed above; then, on larger spatial scales, the distribution of matter becomes uniform.[21(p16),68] But recently, there have been some claims that galaxies interact to form structures even larger than the systems of galaxies mentioned above.[38,69,70] If galactic systems form on all spatial scales of the universe rather than there being some maximum size, then the overall distribution of matter in the universe would lack a characteristic spatial scale, similar to fractals, structures that are self-similar across a wide range of spatial scales. The argument is that when it comes to the distribution of matter, the universe may have fractal characteristics. Nonetheless, the universe is "clumpy" across at least an extremely wide range of spatial scales, suggesting that the distribution of matter throughout it can be considered to be at least partially fractal.

Dark Matter

By studying the properties of some of the fundamental particles, and examining the motions of stars on the outskirts of galaxies and light traveling long distances through the universe, we have discovered that the matter made from the particles in the Standard Model—the types of matter familiar to us—turn out to be a very small percentage of all matter in the universe.[39(pp295-296)] There is another kind of matter made from particles much different from those constituting normal matter because they are believed to only have gravitational and/or weak nuclear interactions with the fundamental particles in the Standard Model; this type of matter is referred to as dark matter because it does not interact electromagnetically, enabling it to evade optical detection by not emitting or reflecting light. Because dark matter interacts with regular matter through only a subset of the four fundamental forces, the interaction between the two types of matter is not as complex as those occurring between regular matter. In addition, while dark matter forces may exist, allowing for interactions between dark matter particles, they too appear to not be as complex as the four fundamental forces enabling interactions between regular matter.[45,47] In addition to observing their gravitational influence on the largest physical systems in the universe, physicists also hope to infer the presence of dark matter particles by detecting the conventional particles that they are believed to decay into after being created in the aftermath of high energy particle collisions.[71]

There is believed to be dark matter particles that reveal themselves on the smallest spatiotemporal scales through their interactions with some of the particles involved in weak nuclear interactions — $W^{\pm}$ and Z^0 bosons, and neutrinos. And because they do not interact electromagnetically, they do not interact much at all with ordinary matter.[39(p433),45,71] But recall that the Higgs boson also has a close connection to the $W^{\pm}$ and Z^0 bosons because it accounts for their unexpectedly high mass. In fact, the Higgs field interacts with all generations of fundamental particles in the Standard Model and is responsible for their mass as well — the more the particles interact with the Higgs field, the more mass they acquire. The properties of the Higgs boson — its limited interaction with the particles in the Standard Model other than the mass it conveys, and its close association to the weak nuclear force — gives it properties similar to dark matter but it is typically not viewed as being a viable candidate for a dark matter particle. However, it has recently been proposed that one of the by-products of a decaying Higgs boson, besides the Standard Model ones that have already allowed scientists to detect the Higgs boson at the LHC, may in fact be some form of dark matter particle.[72]

Dark matter reveals itself on the largest spatial scales through its gravitational interactions with regular matter and it is believed that there is at least five times more dark matter of this type than regular matter in today's universe.[45,73] It does not clump like regular matter to form stars so, therefore, it does not give off light,[39(p295),45] and the only indication of its presence is its gravitational warping of spacetime that affects the motions of both stars at the edges of galaxies and light traveling long distances through the universe. The distribution of this dark matter throughout the universe is nearly identical to the distribution of conventional matter, and generally, this dark matter surrounds the largest structures — galaxies and systems of galaxies — and provides them the gravitational scaffolding that shapes and molds the universe's large-scale structure, fittingly referred to as the cosmic web[45] (Figure 4).

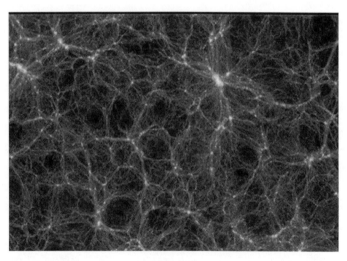

Figure 4

Dark Energy

Remember that in relativity theory, the fabric of spacetime is a dynamic entity capable of warping and stretching in various ways determined by gravitational fields and the motions of objects through space. In addition to this, the universe appears to be suffused with a uniform distribution of energy, called dark energy, whose effect is to cause the universe to expand.[21(p129),69,74] This dark energy becomes dominant in the voids between the largest structures in the universe—galactic systems such as clusters, walls, filaments, etc.—because the field strengths of the fundamental forces through which matter interacts are negligible in these locations.[39(pp238,275),75] The force of gravity is strong enough to counter the outward swell of spacetime and hold galaxies and systems of galaxies together, as well as the solar and planetary systems that exist within them, while the electromagnetic and nuclear forces become dominant on smaller spatial scales and have no trouble keeping fundamental particles together to form a single physical system despite the ever present impulse for space to swell. Therefore, dark energy is only strong enough to expand space in the voids between the largest physical subsystems of the universe, where only virtual particles blink into, and out of, existence.

Many particle physicists are beginning to believe that the Higgs boson is the key to understanding dark matter. It turns out that the Higgs boson may also be critical for understanding dark energy. This is because removing the Higgs field from a region is equivalent to adding energy to that region. The process is said by Brian Greene[39(p260)] to be analogous to noise-cancelling headphones that produce sound waves to cancel the ambient sound entering the headphones. This reduces the background noise and allows you to better hear what you are playing through the headphones. Similarly, cold empty space—the vacuum—is believed to be at its emptiest when the Higgs field is present. Therefore, in addition to pairs of virtual particles randomly produced by quantum fluctuations, the vacuum of space also contains the Higgs field. But if no matter exists in the large voids between the largest structures in the universe, is there still a Higgs field there? If dark energy is equated with the noise that gets suppressed by the Higgs field, and if voids are regions without a Higgs field, then the voids will be noisier regions of spacetime and will expand. The noise that is unfiltered by the Higgs field can be regarded as energy and pressure in spacetime itself that results from more turbulent quantum fluctuations.

Big Bang

The expansion of the universe occurs uniformly because dark energy is uniformly distributed throughout the universe. Therefore, no matter your location, you will see that the universe is expanding in all directions.[21(pp66-67),39(pp230-231)] Furthermore, the farther you look in space, and hence, the farther back in time, the faster the expansion occurs because the effect of dark energy is cumulative.[39(pp278-279)] So no matter your perspective within the universe, the apparent motion of galaxies away from you due to the accelerated swelling of space makes it appear as though you are at the center of an explosion, and that at some point in the distant past, all the matter was once much closer together. This observation has led to the Big Bang Theory, which says that for some unknown reason, our universe began with a coherent and explosive introduction of matter and energy that has resulted in the current expanding cosmic web of matter we see today.[63,76]

If the universe is spatially infinite, then instead of viewing the Big Bang as occurring at one single point in space and the universe expanding out from there, we can view it as occurring everywhere in space simultaneously so that in the right locations, such as the space in between the largest structures in the universe, the density of matter reduces over time, allowing that space to expand.[39(p249)] Thanks to this universal expansion, we are able to determine the age of the universe. In general, the largest galactic structures are considered stationary relative to each other except for their motion that results from being dragged along by the expansion of space. To gauge time, we must observe change in some physical system; for the time since the Big Bang, we use the mass density of the universe because it has steadily decreased thanks to the universe's expansion. This decrease has been uniform, allowing us to use this quantity as a marker for the amount of time that has passed since the Big Bang, which we have measured to be almost 14 billion years ago.[21(p26),39(pp233,235)] And as opposed to being constant, we have found that the rate of universal expansion has been variable throughout the history of the universe. Data suggests that for the first 7 billion years after the Big Bang, the expansion rate was decreasing due to the attractive pull of gravity produced by the initial high density of matter in the universe. Then, as the density dropped and the voids appeared, dark energy began to get the upper hand on the attractive pull of gravity and the universe entered the current period of accelerated expansion.[39(p300)]

When we measure the afterglow of the Big Bang by imaging the most distant regions of the universe, we see that there is a largely uniform distribution of temperatures for the galactic systems in the early universe. This is even true for regions separated by distances so large that light has yet to have time to complete the trip from one region to the other, meaning that they've been completely isolated from each other ever since the Big Bang. But if there are regions of the universe that have not had time to interact at all, how could the entire early universe have roughly the same temperature all throughout?[21(pp43-45)]

Inflation is a cosmological framework for explaining this feature of the universe and it has at least some observational support. In this view, the universe expanded by large amounts in a brief instant soon after the Big Bang and the expansion was mediated by a field whose quanta are called inflatons—in fact, this inflaton field is considered to be a type of Higgs field different from the electroweak and grand unified types.[21(pp51,53),39(pp281-285)] In the short period of time after the Big Bang but before inflationary expansion, the matter of the universe were close enough to interact and come to nearly the same temperature.[21(p45)] During the brief period of inflationary expansion, all of this matter was spread almost perfectly uniform throughout the universe. The matter that was in contact before Inflation separated from each other faster than the speed of light during Inflation but it is important to note that this matter did not travel through space faster than the speed of light—it was the swelling of space itself that broke the cosmic speed limit, a phenomenon that is not outlawed by special relativity.[21(p44)] This brief period of unimaginably fast expansion was over as soon as the inflaton released all of its energy into the production of matter and radiation. After this, the story is the usual Big Bang one,[39(pp281-285)] but while Inflation did result in all regions of the universe having nearly the same density of matter, there were, however, slight nonuniformities due to quantum fluctuations. These nonuniformities have become the seeds for galactic structures because they became the denser regions where gravity caused matter to clump.[21(pp60-61),39(pp305-308)] During this time, dark energy's influence became smaller because gravity was working against it but eventually, the density of matter decreased enough to create voids in between the largest structures in the universe. In these voids, gravitational fields are weak and dark energy has the upper hand and space swells.

Inflationary cosmology ensures that the universe starts out in a low entropy state—the nearly homogeneous distribution of matter and temperatures is a low entropy state when gravitational fields are significant. Recall that entropy is a measure of how likely a system's configuration is—unlikely states have low entropy while states that are more probable have high entropy. Since gravity causes matter to clump, the nearly perfect homogeneous distribution of matter just after Inflation is an unlikely state for the universe to be in, and hence, a low entropy state. Ever since the end of inflationary expansion, gravity has been working to increase the entropy of the large-scale universe,[39(p317)] and the forces currently in the Standard Model have increased the entropy of the universe on smaller scales. This is the Second Law of Thermodynamics and it is actually statistical in nature because it says that physical systems *tend* to evolve towards states of higher entropy, but on small spatial and temporal scales, entropy can momentarily decrease—an example being the quantum fluctuations that temporarily result in pairs of virtual particles. For classical systems, however, there is a strong statistical likelihood that entropy will increase,[21(pp244-245),39(p158),48(pp51-52)] and work has to be performed on a system in order to decrease its entropy or to halt its march towards higher states of entropy. The by-product of performing this work is a greater amount of entropy released into the environment. Therefore, on large spatiotemporal scales, entropy is always increasing, a phenomenon responsible for the observed direction to the flow of time.

Theory also Expands the Universe

On the highest levels of the spatiotemporal hierarchy, matter forms an expanding and interconnected system of galaxies resembling a cosmic web. Dark matter encapsulates this web and provides it the gravitational scaffolding necessary for it to form. The expansion of spacetime occurs in the voids between the largest structures because they are the locations where the four fundamental forces are negligible and the uniformly distributed dark energy becomes dominant.

We saw that the Big Bang occurred almost 14 billion years ago and put the universe in a low entropy state. Immediately after the Big Bang, matter and energy was, for a brief moment, in a very dense state where there were interactions that resulted in an almost uniform temperature distribution. Then in a sudden universal burst, matter and energy was spread almost uniformly in such a way that matter close to each other before Inflation was rushed apart faster than the speed of light due to the expansion of space. The small quantum fluctuations in the dense, pre-inflationary universe resulted in small differences in the concentration of matter and ultimately became the seeds for galaxies thanks to the force of gravity which, ever since the Big Bang, has been working on large scales to increase the entropy of the universe, something we see as a manifestation of the Second Law of Thermodynamics.

This chapter provides the foundation that you will need to be able to see the correspondence between the universe birthed by the Big Bang and a particular information processing event that occurs in our brains when we sleep at night.

It turns out that dark matter and dark energy are not the only ways in which we can bring ourselves to expand our concept of what exists in the universe. A developing theory called M-theory, viewed by many as the best candidate within modern physics to explain the fundamental nature of physical reality, has done this by including new objects, both small and large, and extra dimensions of spacetime. In the next chapter, we will discuss string/M-theory and introduce a model of the universe that it can be used to construct and we will see how it all relates to what we've considered thus far about the nature of reality.

Chapter 6

Holographic Universe

A Theory-of-Everything

The holy grail of physics is to discover one single theory that is capable of explaining the fundamental nature of physical reality, a reality that consists of a wide variety of interacting systems with sizes ranging from the smallest elementary particle to the entire universe itself. Quantum mechanics does a marvelous job explaining small, low energy, and isolated systems, and when we use it to describe larger, more classical systems, we find that it joins smoothly with the Newtonian description. At the opposite extreme, when classical systems are moving at speeds close to the speed of light, or when they produce intense gravitational fields, we use Einstein's relativity theories—both special and general—to accurately describe the situations. Together, quantum mechanics and relativity theory explain an enormous range of the phenomena that we know to exist but we find that we cannot completely unite them both when we need them to explain the same system, i.e., when the system under consideration is both extremely tiny and extremely massive; objects that fit this description include black holes and the very early universe just after the Big Bang. For these systems, the two theories do not appear to be compatible, something that is evidenced by the lack of the graviton in the quantum-based Standard Model of Particle Physics which does, however, manage to model the electromagnetic force and the two nuclear forces.[21(pp77-78),39(pp337-338),63,77] Therefore, in order to obtain a "theory-of-everything", we will have to find one that is capable of matching the successes of both quantum mechanics and general relativity in explaining their respective aspects of the universe, and is also capable of going beyond them to

81

explain the phenomena that they fall short on, namely, black holes and the Big Bang. Being that the Standard Model already accounts for three out of the four fundamental forces, it is reasonable to suspect that the theory-of-everything will provide a quantum description of gravity that would fit right alongside the quantum description of the other three.[39(p329)]

String Symphony

Recall that in quantum mechanics, the fundamental particles are modeled as structureless, point-like regions of spacetime where there is a certain amount of energy manifesting as mass and motion. Furthermore, these particles are either matter particles communicating via the four fundamental forces or are force particles that mediate the communication between the matter particles. String theory was born when it was discovered that it may be possible to describe all of these fundamental particles as extremely tiny vibrating entities that resemble strands of energy—strings. But in order for this to happen, the theory says that in addition to the three spatial dimensions that we are aware of, these strings need to vibrate into six additional spatial dimensions. Therefore, including time, string theory says that strings vibrating in a 10-dimensional spacetime may be responsible for creating all of the particles in the Standard Model. It considers the vibrating strings responsible for creating all of these particles to be identical so that it's just the different patterns of vibration that determine exactly what kind of particle gets created.[21(p79),39(pp17-18,346)]

String theory is up to the challenge of combining quantum mechanics and general relativity because it is a mathematically consistent theory capable of modeling the kind of quantum particle that corresponds to the graviton. In addition, it has the potential to be more fundamental than the Standard Model because it can explain particle properties without assuming them as input—properties such as mass, electric charge, spin, etc., are of length and time in M-theory are the same quantities that are in quantum theory, namely, the Planck length — the length of a string — and the Planck time — the time it takes light to travel the length of a string.[39(p350)] Each of these discrete locations has six tiny curled up spatial dimensions that are too tiny for us to perceive either directly, or with our scientific equipment. They are arrayed throughout the three extended dimensions of space and represent locations within the universe where physical processes can occur via string vibrations. In short, M-theory says that the entirety of spacetime is the result of string vibrations within a 10-dimensional brane, and if correct, spacetime is akin to a computation performed by the processes occurring within branes[39(pp486-487)] (Figure 5c). Furthermore, the extra spacetime dimension added by M-theory is the dimension along which two or more distinct branes can interact and form a system of branes via fields that they produce — branes can be warped and threaded with fluxes produced by other branes.[21(pp124-125)]

But before we can use M-theory to derive all particle properties mathematically, we must first understand the nature of the extra dimensions that strings vibrate in. A spatial dimension is any independent direction in which you could move, such as the three dimensions of space that we perceive. Any motion within these three dimensions can be expressed as some combination of motion along the three independent directions. The first burning question is: if in M-theory there are extra spatial dimensions, how come we only perceive the three that we do? We just saw one answer which is that six of these unseen dimensions are tiny and curled up at every point in space, too small for us to directly perceive or use scientific equipment to do so.[39(pp362-365)] Another possible reason we do not see these extra dimensions is that the electromagnetic force may be confined to the three extended dimensions that we do perceive — if photons can't leave these dimensions to interact with entities on other dimensions, then we can never see those entities.[39(pp392-393)]

The exact nature of the small curled up dimensions is very important since it is string vibrations within these dimensions that determine the ingredients of the Standard Model—particle mass, electric charge, spin, etc. The equations of M-theory still do not provide additional details regarding the exact shapes of these curled up dimensions which are defined by fluxes—twists and turns that energy follows, leading from one dimension to another.[16,21(p90),39(pp371-372,383)] But by analyzing the equations of M-theory that have managed to emerge thus far, string theorists have been able to limit the possible shapes to a particular class called Calabi-Yau shapes. The problem is that there are numerous Calabi-Yau shapes that are equally as valid and there is no known way to discern which shape, if any, corresponds to the geometry of our universe. The exact nature of the particles created by the different Calabi-Yau shapes will be different since they each have a different geometry. This is a limiting factor to M-theory's ability to make definitive predictions,[21(pp90-91,162),39(p372)] therefore, the hunt is on to identify the particular Calabi-Yau shape that defines the discrete pixels of the universe that we inhabit.[21(pp162-163)]

Despite M-theory being incomplete, we are still able to infer some details about how strings vibrate through the geometry of a particular Calabi-Yau shape and through a system of branes to create some of the particle properties in the Standard Model. For instance, strings come in two varieties: snippets and loops[21(pp116-117)] (Figure 5a-b). In general, two strings interact by coming together when their endpoints fuse; the alternative interaction is for strings to break apart. These interactions allow strings to grow, and for snippets to transform into loops and vice versa.[79] The endpoints of snippets are confined to a particular dimension but loops are not; they can fly off all dimensions within a brane and can be viewed as two snippets with their ends fused together. The graviton is believed to be a loop while the other force-mediating particles are believed to be snippets. Therefore gravity is the lone force believed to be able to interact through all of the spatial dimensions while the rest are believed to be capable of interacting with only a subset; in the case of electromagnetism, this subset is the three extended dimensions that we are directly aware of.[21(pp116-118),39(pp392,394)] This may also provide an explanation for the weakness of the force of gravity relative to the other three forces—gravity gets spread throughout more dimensions. In addition, recall that we do not have a good explanation for the existence of exactly three generations of fundamental matter particles in the Standard Model. Detailed information regarding the fluxes through the dimensions in Calabi-Yau shapes could provide the answer. Specifically, the fluxes between the dimensions in a Calabi-Yau shape can encode topological "holes", or "cavities", that essentially allow for the existence of different classes of particles. Because the Standard Model says that there are three such classes, or, generations of particles, most string theorists believe that the Calabi-Yau shape of our universe encodes three holes.[16,39(p373)]

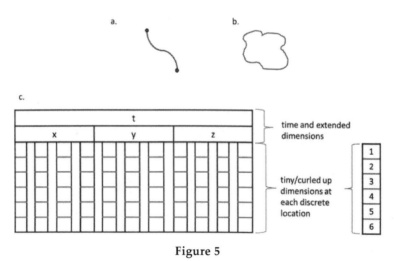

Figure 5

can be connected through the 11th dimension to another brane that can differ in terms of the number of dimensions and the geometry of its Calabi-Yau shape. String theorists have begun to develop models of the universe using this particular framework; one such model is the Holographic Universe. In this model, activity on a boundary brane, or holographic screen, leads to a projection onto another brane known as the interior, resulting in the physical universe that we directly perceive (Figure 6).

A version of the Holographic Universe has been found where a four-dimensional, gravity-less, and point-particle quantum field theory existing on the boundary brane projects to a 10-dimensional string theory with gravity in the interior brane in such a way that the boundary encodes all of the information of the Interior.[21(p268)] Although the processes on the two branes look different, they represent the same physics so that the evolution of the two branes is tightly interlocked[21(p261)]—anything that happens in one brane has a corresponding event that occurs in the other, making them, in a sense, a form of parallel universe. In this model, it is possible that a single particle in the interior—the physical universe—could correspond to a collection of particles that are considered to represent one entity on the boundary, but since the two are so tightly interlocked, the probabilities for corresponding events are the same in both realms.[77] The boundary of the interior universe contains particles that are similar to those in the Standard Model and they interact in a way that is very similar. For instance, the boundary contains particles that behave similar to how quarks and gluons interact in the interior, but on the boundary, the interaction is more complex because there are more than three color charges.[77] Another distinguishing characteristic between the boundary and interior branes in the Holographic Universe model is that although gravity does not act in the boundary, it emerges from processes there and projects to the interior and moves masses around there in such a way that entropy increases—quantum fields on the boundary brane produce gravity on the interior brane.[21(p272),59,77]

Because M-theory is still in its developmental stages, any model resulting from it, such as the Holographic Universe, is only speculative. And just like we do not know how to single out one of the possible geometries for the extra dimensions that pertains to our universe, we also do not have a boundary theory that results in an interior theory that matches our universe.[77] If string theory also applies to the boundary brane, then one might suspect that the point particles of the quantum field theory on this brane may also be represented by strings vibrating in 10 spacetime dimensions; in this case, it is perhaps a 10-dimensional boundary brane that projects to a 10-dimensional interior along a dimension joining the two (Figure 6). Nonetheless, as Brian Greene suggests, holographic ideas have stood up to the scrutiny of many experts in the field and are now a popular way for them to view the universe,[21(pp269,272)] and one of their main tasks is to generalize the results so that they apply to a universe like our own.[21(p273),39(p485)]

Vibrating Strings, Quantum Systems, and Gravity?

From the insights of quantum mechanics, we know that all particles are associated with nonphysical probability waves that tell us the likelihood the particles will manifest one of their possible configurations when measured. To this day, we are still debating the reality of these mathematical functions—do they exist somewhere in reality or are they just an abstraction? Furthermore, recall that quantum mechanics says that the universe is nonlocal, meaning, the properties of two quantum systems can become correlated over distances as large as the universe so that when one system is measured and its wave function collapses, the other system's wave function will simultaneously collapse in a coordinated fashion. I speculate that the holographic screen, or, boundary brane of the Holographic Universe is a potential location for quantum probability waves to exist, and because in this case they exist only on the boundary brane, they would not be considered to be physical entities—for any system to be considered physical, it must exist in the interior brane. In addition, nonlocal holographic data storage on the boundary brane, the brane that projects the physical universe, would be consistent with the nonlocal nature of the universe. In Figure 6, the nonlocal nature of the boundary brane is represented by a lack of division in the top row between the four extended dimensions of spacetime, the dimensions that we directly perceive. This is contrary to the representation of these four extended spacetime dimensions in the top rows of the interior brane, where once manifest, physical processes are local.

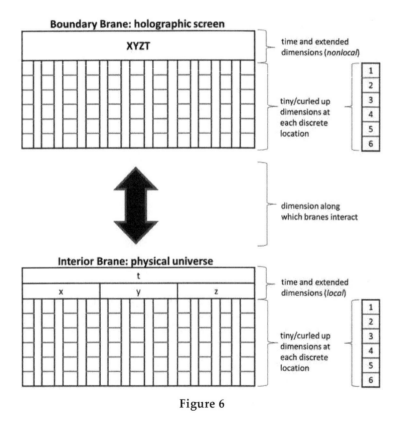

Figure 6

Interestingly, it has recently been found that quantum mechanics can be "derived" from a string/M-theoretical framework.[80] If this is true, then it is evidence in support of my suggestion that perhaps the four-dimensional quantum field theory on the boundary brane of our most concrete Holographic Universe formulation to date can somehow also be expressed in terms of the 10-dimensional spacetime of string theory. Or more simply, I am suggesting that the long-awaited final version of a complete M-theoretical description of the universe—the so-called "theory of everything"—might be a 10-dimensional boundary brane projecting, through the 11th spatial dimension, to a 10-dimensional interior brane.

We also saw earlier that, increasingly, gravity is being looked at as the product of processes occurring within matter related to information and entropy. In addition, we saw that black holes are essentially just pure gravity and display holographic characteristics.[16] In the Holographic Universe, information stored on the boundary brane—the holographic screen of the physical universe—is the source of the entropy resulting in gravity in the interior.[59] Therefore, for every object with mass in the interior, there are processes occurring on the boundary that result in the object's gravitational field. In this view, a black hole isn't just a singularity and an intense gravitational field, it also corresponds to a very large swarm of interacting particles on the boundary brane. Since gravity is not present on this boundary, the conflict between quantum mechanics and general relativity that arises when we attempt to use them both to describe black holes is avoided. As is usually the case with holograms and their projections, the two manifestations—large swarm of particles on the holographic boundary and pure gravity in the interior—may look nothing alike but they still embody the same information.[21(p269),77] Therefore, the Holographic Universe model appears capable of resolving the differences between quantum mechanics and general relativity when it comes to explaining black holes.

Cyclic Universe

It turns out that the Holographic Universe is compatible with another M-theoretical model, one that describes the Big Bang, called the Cyclic Universe. This model says that Big Bangs occur when two branes that can interact through the 11th dimension collide periodically; the two branes cycle through periods of being attracted to each other, which ultimately leads to collisions and the beginning of a new universal cycle.[21(p123),39(pp406-411),63] But recall that Inflation is a concept that is used to explain the flatness of spacetime and the homogeneity of temperatures, and hence, the density of matter, throughout the early universe. The Cyclic Model is also consistent with Inflation because the branes are already flat, and when they collide, they hit simultaneously everywhere to produce an almost uniform distribution of matter and energy. Furthermore, quantum fluctuations on the branes just before the collision are all that is needed to provide enough variation in the distribution of matter throughout the universe to seed future galactic structures.[81] If the Cyclic Universe model is accurate, our universe could be one of an ensemble of universes that come into existence one after another.[21(p120),39(p412),82]

Supersymmetry

Recall that cosmological evidence suggests that our universe has a dark side, namely, the ubiquitous and mysterious universe-swelling force called dark energy, and another type of matter called dark matter that is currently not accounted for by the Standard Model of Particle Physics. A close examination of M-theory reveals that it too hints at nearly undetectable matter in the form of particles that are complementary to the ones currently existing in the Standard Model. If these particles exist, their complementarity to the Standard Model would give the universe a symmetric quality. Fittingly, the theory that posits these particles has been termed Supersymmetry and many theoretical physicists believe it represents the best candidate theory to extend the Standard Model.[83,84]

The two classifications of fundamental particles in the Standard Model are: matter particles called fermions, and force particles called bosons. Supersymmetry says that all of the particles in one category have extremely massive partners that belong in the other category. This means that each fermion in the Standard Model has a much heavier boson partner particle; the same goes for the bosons in the Standard Model—they have extremely massive fermion partners. So far, however, none of these "superpartners" have been detected in the world's particle accelerators, most notably at the Large Hadron Collider (LHC) where the Higgs boson was recently detected. However, some of the predicted masses for these particles require that the LHC be operated at higher energies before they can be detected. In fact, the LHC has recently undergone an upgrade that will allow it to smash particles together with enough energy to better probe the limits of the Standard Model; i.e., the collider's power will increase so that it may become possible that exotic forms of particles, such as the superpartners of Supersymmetry, and maybe even other types of Higgs boson interactions, may be produced.[83,84,85,86]

Supersymmetry was proposed to address a problem: we expect the Higgs boson to be much more massive than it has been predicted and found to be. Supersymmetry says that the superpartner particles cancel some of the mass that the Higgs boson could acquire from its interactions with Standard Model particles via the fundamental forces. In addition, the superpartners in Supersymmetry are viable candidates for a form of dark matter because they are believed to interact weakly with normal matter, and if they are found, they will probably have masses within the proper range to produce the observed gravitational influence of dark matter throughout the universe.[83,84]

Observable Problems

String/M-theory initially began rooted in aspects of physical reality that we could make observations of; it was conceived in an attempt to explain the particles in the Standard Model but eventually grew so much and so fast that it outpaced our ability to observe its predictions. Two consequences of this are that: 1) the theory is unfalsifiable because we cannot test it—something that is considered essential for it to be looked at as a viable scientific theory; 2) we do our best to describe the mathematical concepts predicted by the theory, but we have no solid interpretation for what they actually mean.[21(p171),39(p376),87]

All hope is not lost though because quantum mechanics is an example where we have a very successful theory, but yet, not every mathematical concept of the theory is observable. It shows us that there just needs to be enough predictions that can be verified through the observations that we can make.[21(pp169-170)] In the case of M-theory though, the possibility of observing a string is slim-to-none, but we can still try to find a physical system that we can observe which models the universe in some way sufficiently enough for testing string/M-theoretical predictions. Observations of this system may provide indirect evidence in favor of, or against, M-theory. One way to obtain a model might be to program a computer—most likely a supercomputer, or perhaps one day a quantum one—to simulate the processes described by M-theory.[88,89,90] Another potential model of an M-theoretical universe is a physical one, a particular type of crystal that undergoes processes analogous to the predictions made by M-theory regarding large-scale signatures of strings that are remnants of the very early universe.[22,91] In the end though, no matter how satisfying M-theory may be, until direct, or indirect, experimental observations are obtained that will allow for the theory's predictions to be proved or disproved, it will forever remain an elaborate, mathematical/conceptual system with no known physical analog.[21(p171),87]

Mathematical Bloodhounds

The emergence of M-theory has forced us to reconsider both what the fundamental constituents of matter are, and what the limits of spacetime are. We found that the point particles of the Standard Model can be represented by tiny vibrating strands of energy called strings, but these strings need to vibrate into an additional six tiny and curled up dimensions at every discrete chunk of spacetime. Together, these six tiny curled up dimensions, the three extended dimensions that we directly experience, and the dimension of time define a 10-dimensional entity called a brane. In addition, this brane can be connected through an 11th dimension to another brane that may or may not have the same number and shape of dimensions, making the universe a product of processes, namely, string vibrations, occurring within a system of branes. Theoretical physicists are using this framework to develop models that explain our universe, such as the Holographic Universe, the Cyclic Universe, and Supersymmetry. Models like these turn out to be compatible with our more established views of the universe—such as quantum mechanics, general relativity, and Inflation—and have shown an ability to provide explanations for phenomena that these theories cannot.

This chapter prepares you for catching a clear glimpse of the similarities that exist between a universe described by string/M-theory, and the detailed neuronal circuitry within key areas of the brain.

As impressive as this incomplete theory is, M-theory may be unfalsifiable because there appears to be no hope of directly observing many of its predictions. We are trying to follow its mathematics to their logical conclusion, but find that there are many equally valid ways that the universe could be within an M-theoretical framework. To circumvent our inability to directly measure strings and shed light on the many different possibilities, we must look for ways to model a universe described by M-theory using some type of physical model and super/quantum computer simulations.

Before proceeding to a discussion of the system that I assert to be a model of the universe, namely the human brain, we will revisit another important concept introduced in Figure 1—complexity. In the next chapter, we will consider the evolution of complexity throughout the universe.

Chapter 7

Universal Evolution

Complexity and Evolution

The concept of evolution is most often used to refer to changes that occur to living species genetically, and hence, physically as well, over many generations as a result of the long-term interactions they have with their environment. However, even though nonliving physical systems do not have genes, their states—internal and external configurations—also change due to their interactions with their environment, interactions that are ultimately governed by the fundamental forces of nature. Because the universe consists of a hierarchy of physical systems—both living and nonliving—that exist across a widely varying range of spatiotemporal and complexity scales, where systems and their environments engage in a never-ending dance of adjustments to each other, the universe as a whole changes in time and can be considered to be evolving as well. Therefore, it turns out that we can consider biological evolution to be part of a much broader concept of evolution, one that includes the whole universe.[37(p181),92(p27)] But thus far in our discussion, we have mostly considered the universe across its vast array of spatiotemporal scales and largely neglected the fact that physical systems also vary across many levels of complexity. In this chapter, we will consider the evolution of the universe and the complexity thresholds that get crossed by the physical systems appearing throughout this process.

To accomplish this, I will use the Holographic Universe framework, and describe its evolution using established laws of physics as well as some of the leading theories currently under development. Recall that in this framework, a system of two branes computes physical reality, a reality that we directly perceive to be four-dimensional, but is actually eleven because, along with the four extended dimensions of our everyday world, there are six tiny curled up ones, and another that connects the two branes so they can form a single system. The boundary brane is a holographic screen within which there are processes that project the physical universe—matter and gravity—to the interior brane, which is where the spatiotemporal-complexity hierarchy of Figure 1 resides. Therefore, since the system of branes in the Holographic Universe framework also has a hierarchical configuration, the physical universe is, in fact, a hierarchy existing on one level of another hierarchy—in terms of hierarchy theory, the physical universe that we perceive, and measure with our scientific equipment, is a "heterarchy" existing in the interior of the holographic system of branes[37(pp65-66)] (Figure 7).

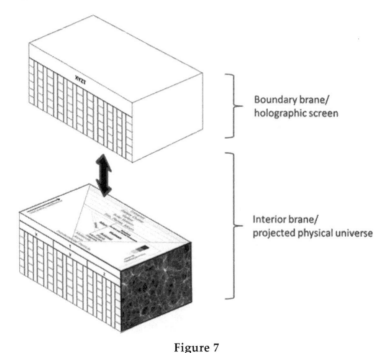

Figure 7

The course produced by The Teaching Company entitled *Big History* provides an excellent framework for considering the complexity thresholds that are crossed throughout the evolution of the universe.[49(pp7-8)] In order of increasing complexity, the eight thresholds that it defines are: 1) the creation of the universe during the Big Bang, 2) the formation of stars, 3) the formation of atomic elements, 4) the formation of planets, 5) the appearance of life, 6) the appearance of humans, 7) the development of agriculture, and 8) the development of modern society. These complexity thresholds are crossed in localized regions of the universe and the rate at which they are reached here on Earth has been increasing.[49(p12)] This does not mean that there is a constant forward march toward ever increasing levels of complexity, rather, it means that higher levels of complexity come about when the conditions are right, and sometimes, in an adjustment, a system can fall to a lower level of complexity.[37(p127),93] In general, as long as conditions are favorable, such as there being the right ingredients and sufficient sources of energy in a particular region of the universe, physical systems with more and more complex forms can arise.[49(pp15-17,54-55)] This "complexification" that has occurred in our local region of the universe has resulted in the spatiotemporal-complexity hierarchy presented in Figure 1, as well as other forms of stratification that arise on finer scales, some of which we'll encounter later in this chapter.[5(Ch7),94]

Prebiological Evolution: Complexity Thresholds 1, 2, 3, and 4

Because I am using an M-theoretical framework to describe the universe, I can consider the Big Bang to occur when two or more branes "collide", an event that marks the beginning of a new universal cycle and is considered as the **1st complexity threshold**. Recall that within a Cyclic Universe framework, the attraction felt between two branes could occur periodically so that universes are created one after the other. Since both the

Cyclic and the Holographic Universe models involve systems of branes, it is possible that they may be compatible with each other such that the two branes in the Holographic Universe are the ones that collide cyclically, or on the other hand, there may be some other brane(s) out there that collides with one or both of these branes. Nevertheless, it is believed that such a collision occurs only once, at the very beginning of a universal cycle, and after it happens, the universe is a closed system so that energy is no longer introduced into it. The energy produced during the Big Bang can, however, change forms, and does so in such a way that work is performed on physical systems, a process that converts the energy into a less useful form and increases the net entropy of the universe.[49(pp15-17)]

The colliding branes cause strings to vibrate within them, creating first the Higgs field, or, "Higgs ocean", then the fundamental particles in the Standard Model. The activity on the boundary of the universe is highly correlated with the interior and projects the physical universe there. In addition, the collision between the branes occurred in such a way that the subsequent string vibrations produced a distribution of matter and energy that was nearly uniform, and quantum fluctuations are believed to have accounted for the tiny differences that existed within this distribution throughout the early universe. These tiny differences later became the seeds of galaxies and systems of galaxies.

Immediately after the Big Bang, all that existed were the fundamental particles in the Standard Model and their antiparticles.[40(p9),42] Then, all of the higher generation matter particles, who require high energy densities to be created, decayed into the first generation particles—the ones that make up ordinary matter. In addition, the still somewhat mysterious asymmetry between matter and antimatter kicked in to eliminate most of the antimatter created during the highly energetic Big Bang. Eventually, as temperatures decreased, quarks began to combine to form nucleons, nucleons combined to form nuclei, and nuclei combined with electrons to form the most basic of our atomic elements, hydrogen and helium.[40(p6),49(p33),65,74]

The nearly uniform distribution of matter in the early universe was a low entropy state since gravity is a force that causes matter to clump, and it set the stage for one of the underlying themes of cosmic evolution—the Second Law of Thermodynamics, which states that systems tend to occupy states of configuration that are more probable.[49(pp15-16)] From this initial low entropy configuration, the physical processes occurring on every spatiotemporal scale have resulted in a net increase in the total entropy of the universe over time.

Thanks to gravity, the slight deviations in the nearly uniform distribution of matter became the seeds of galactic structures because it caused the hydrogen and helium that formed in the aftermath of the Big Bang to clump and form stars—structures that propel the universe across the 2nd **complexity threshold**. These stars are the elementary parts of galaxies which are themselves the elementary parts of systems of galaxies such as filaments, clusters, and walls. Together, these structures, along with dark matter, form the cosmic web of the large-scale structure of the universe. Stars are distributed all throughout this cosmic web, and inside each one, the 3rd **complexity threshold** is crossed when hydrogen and helium are fused together to form heavier and more complex atomic elements, a process that releases heat and electromagnetic radiation into the star's environment.[39(p354)] Through the fusion that occurs inside them and when they explode, stars create all of the elements in the periodic table and provide energy to the rest of the universe.[40(pp6,41)]

When stars die, they cease to release heat and electromagnetic radiation into their environment, but they do leave one last vital ingredient that helps continue the increase in complexity throughout local regions of the universe—they scatter the atomic elements that were forged inside them into their surroundings.[49(p35)] And once again, gravity acts on this matter and causes it to clump and form the ingredients of solar systems such as another star, or planets, moons, comets, asteroids, etc.[40(pp9,41),49(p40)] This marks the passing of the **4th complexity threshold**, and the chemical composition of these physical systems is ultimately determined by the composition of the clouds of stellar remnants that they formed out of. Solar systems are formed when the smaller objects are attracted into orbit around stars, which situates them somewhere within the environment of the star so that they are exposed to the energy that it radiates. The cloud that our solar system formed out of was believed to consist of: 70% hydrogen, 27% helium, approximately 1.5% carbon, oxygen, and nitrogen (all vital ingredients of living organisms), and the remaining 1.5% included all the other natural elements in the periodic table.[49(p41)] When Earth formed, a process called differentiation caused the heavier elements—metals—to sink to the core while the lighter elements stayed on the surface, such as those comprising bodies of water and the atmosphere.[49(pp45-47)]

Biological Evolution: Complexity Threshold 5

The Earth is at least one location in the universe, but there are most likely many more, where conditions suitable for the appearance of life exist, such as the presence of ideal ingredients and ambient energy levels that are not too hot like the surface of a star and not too cold like the vacuum of space.[49(pp68-70)] Earth is a planet made from a wide variety of atomic elements, each with unique properties, and is positioned in an orbit that enables it to be bathed in a usable amount of the sun's energy. Furthermore, it is still a highly dynamic physical structure, all throughout, thanks to its molten metal core, shifting tectonic plates, flowing bodies of water, and dynamic atmosphere. Together, these sources of energy are used by special ingredients to interact and form life forms, types of systems that possess enough emergent properties to cross the 5th **complexity threshold**. These special ingredients interact to perform the work that is necessary for living systems to maintain their level of complexity and evade the advance of entropy, consistent with the Second Law of Thermodynamics. But recall that this process does indeed have a waste product, one that ultimately increases the entropy of the environment more than the system decreases its own, and it comes in the form of less useful energy. Therefore, the collective action of all life on Earth is to down-convert the ambient energy into a less useful form.[48(pp104-106)]

Life forms are complex systems consisting of diverse components that each perform specialized functions and are bound into larger structures that exhibit the types of emergent properties unique to life. One of those properties is metabolism, which is the process by which life forms take in chemicals from their environment and build them into the products they need to live. This is a necessary process if all of the elementary units of organisms are to successfully bind together into a single stable system capable of temporarily warding off decay by importing energy to do work and decrease entropy.[95] Contrast this with the case of a nonliving system whose entropy will increase if no work is performed on it, causing it to eventually reach thermodynamic equilibrium with its environment. The right ingredients, such as water, the surfaces of rocks, CO_2, and thermal energy, make metabolism inevitable, leading initially to the formation of simple, single-celled organisms. These conditions existed early on in Earth's history and probably appeared first underwater, near hydrothermal vents because the molten and toxic conditions of the early Earth's surface were not conducive to life.[95,96]

Organic compounds, the vital ingredients of life, result from the interaction of carbon, nitrogen, oxygen, and hydrogen, all of which are abundant on Earth. Life here is carbon-based, which means that the carbon atom plays a central role in the formation of organic compounds because its versatility allows it to bind stably with up to four atoms. Such flexibility means that carbon interactions can be at the heart of an extremely wide variety of physical forms.[95] More complex life—multicellular organisms—require that simple cells cooperate to form more complex systems of cells. A crucial step in life's advance to ever more complex forms was the acquisition of prolific producers of energy—mitochondria—by complex cells. This is important because more energy is required to support higher levels of complexity, therefore, the energy produced by mitochondria made it possible for more complex life forms to evolve.[95,96]

Two more emergent properties of living systems that distinguish them from nonliving ones are reproduction and adaptation. These properties are important because the environment of a living system has a profound impact on the trajectory that its evolution takes, and to survive, these systems must either adapt to the environment or change it. Furthermore, these reactions must happen at least as fast as the change happens or else the organism risks going extinct.[37(pp103-104)]

The first life forms that appeared on Earth—simple single-celled organisms—reproduced through some type of cloning, which does lead to multiple generations of the species but with little variation from one generation to the next, and consequently, species could not adapt much to changes in their environment. On the other hand, one of the special subcomponents of more complex life is a molecule called DNA that provides the genetic material responsible for species' abilities to adapt much faster to their environment, particularly when they reproduce sexually, producing greater variation in offspring because they now contain a mix of genetic material from two parents.[49(p74)] In addition to this combinatorial source of variation, offspring also vary because genes are never copied with 100% perfection. Both of these sources of variance lead to the organisms of a species having slightly different characteristics, and nature selects the ones that are most fit to survive in their environment, similar to how a pigeon breeder breeds pigeons to his or her specifications.[49(p60)] Over time, this process determines which species will thrive and which ones will struggle within a given environment—those with genes that increase their chances of survival therefore also have a higher chance to live, reproduce, and pass genes on to the next generation. The end result is that DNA stores the detailed information of what evolution has produced along its many adaptive radiations when new organisms appear and rapidly evolve into a wide range of species that vary in complexity, from the simplest single-celled organism, to the most complex organism of all—humans.[49(p77)] It is important to note, however, that the process of natural selection is actually statistical and is about change in the average properties of a species over many generations.[49(p61)]

Ultimately, all life forms on Earth share a common ancestor from which the evolution of species has branched into a multitude of trajectories throughout the history of life. The more recent the split between the trajectories of any two species, the more similar they will be, and hence, the more overlap there is in their genetic information. If one species goes extinct, its genetic information is not lost completely because at least some of it still exists within other species. Therefore, genetic information can survive the death of a single member of a species, as well as the extinction of the entire species itself. In this view, the genetic information within an individual organism is what survives over time and drives reproduction so that the physical form an organism takes—its body—is just a vehicle, one particular manifestation of the underlying genetic information existing within it.[48(pp95,100)]

Recall that life most likely appeared first in the water because, there, conditions conducive to life existed sooner than they did on the molten surface of the Earth, and in addition, the Earth's atmosphere was still toxic. Through natural selection, the first single-celled organisms branched along many trajectories of evolution and produced multi-celled organisms, and then several fish, pre-reptilian, and pre-amphibian animal species. But eventually, after the Earth's surface cooled, hardened, and other hostilities subsided, life eventually spread to land. Insects and plants most likely ascended first, which was a feat that not only marked a major milestone for life because it had reached land, but also because the Earth's atmosphere was transformed in the process since plants harness energy through photosynthesis— the utilization of the abundant CO_2 and sunlight at the Earth's surface to extract energy and produce oxygen as a by-product. This helped further pave the way for the evolution of oxygen-breathing life forms. The result was many adaptive radiations producing several species of reptiles, amphibians, mammals, and eventually the dinosaurs, who ruled the Earth for many millions of years.

But recall that the environment determines the trajectories that life evolves along, i.e., all adaptive radiations are shaped and molded by the environment, and as it changes, a species' need to adjust drives its evolution. This is not much of a problem when the change occurs very slowly over many millions of years as it usually does, but it becomes a problem when change occurs abruptly and jeopardizes the ability of the most affected species to keep pace with the changes. This happened in the case of the dinosaurs, perhaps due to a combination of climate change and a comet impact, causing them to go extinct roughly 65 million years ago. They, and other complex life forms, could not survive this Earth-transforming event, leaving small amphibians, reptiles, and mammals as the most complex life forms on the planet. As the Earth settled near the conclusion of this extinction event, life set out along new evolutionary trajectories to produce the hierarchy of life that we are familiar with today.[49(pp77,81)] Birds joined the list of life forms and a wide variety of species in all animal kingdoms with ever more hominids had brains just a fraction the size of our own. For instance, these early hominid species did in fact make tools and had limited creativity but they made only minimal advances in technology and linguistic ability, consequently, their way of life largely remained stagnant for as long as they existed. Furthermore, despite their more human-like appearance, we probably wouldn't recognize them as being human if we were to actually encounter them in person like our early human ancestors did. After appearing in Africa roughly 250 thousand years ago, the first humans began to spread around the world and exist alongside earlier hominid species, such as Ergaster and Neanderthal, who had already begun using their newfound bipedalism and mental capacity to explore and populate land outside of Africa millions of years earlier. This coexistence would not last to today, however, only up until about 20-30 thousand years ago when species like Ergaster and Neanderthal became extinct, probably under pressure from humans.[49(pp83,94)]

Humans exhibited communication superior to that of any other hominid, including Ergaster and Neanderthals, which has led to their ability to engage in an activity called collective learning—the emergent property possessed by human society that is largely responsible for the universe crossing the **6th complexity threshold**. Before acquiring language, reading, and writing, humans and earlier hominid species only had individual learning, what one organism has acquired through its direct interaction with the environment. This resulted in a very local impact on the beliefs and actions of others, and when the individual died, most of their information was gone with them. In short, language, reading, and writing made it possible to communicate over a broader spatiotemporal expanse and has allowed for a collective wisdom, knowledge, and memory to develop;[37(p263),49(pp90-92),92(p27)] and because of this collective learning, human society evolves at a much faster pace than humans do genetically, and all members of society can contribute to the development of conceptual systems—organizations of ideas expressed in symbolic form—which can sometimes be used to purposefully alter physical objects, thus, creating a bifurcation in the types of nonliving physical systems: natural and manmade. The reason humans are so good at reducing the entropy of their environment by performing work on objects within it is that they use knowledge about physical matter acquired through collective learning[97] which has led to advanced technological creativity and increased adaptability, enabling humans to thrive in a wide range of environments and deal with threats to their way of life. An example of a threat that humans are no stranger to is environmental change. For instance, ice age cycles have averaged 100,000 years in duration and have been separated by 10,000 year interglacial periods of milder climates. So far, humans have used their ingenuity to survive through the harsh conditions of two ice ages and the climatic upheavals associated with the transitional states between them and the interglacials.[49(pp103-104)] In fact, modern humans almost became extinct about 74,000 years ago due to extreme climate change and their numbers may have been down to roughly 10,000 adults of reproductive age. Ultimately, however, it was collective learning that has made it possible for humans to survive and live on, and it remains the essential human characteristic that drives the evolution of society.

In many ways, collective learning leads to an evolving system of knowledge that is analogous to the evolution of life. As we saw earlier, life on Earth is constantly exploring the possibilities of the environment, adapting to it, and when the conditions are right, developing better ways to extract useful resources from it. The result is an ever changing hierarchy of life forms ranging from simple, single-celled organisms, to the complexity of human beings,[49(p56)] and over time, some species go extinct because they cannot keep up with the pace of change, while others evolve to fill survival niches. With the arrival of humans, a similar process is played out in the evolution of beliefs about the nature of reality that have been collectively acquired through human experience, leading to a wide variety of systems-of-thought, such as worldviews, religions, sciences, etc. This too is a process that unfolds over time so that some systems-of-thought fall into- and out-of-favor while others come into existence when new discoveries are made or when existing knowledge is combined in new forms. And just like the relationship between the features of living organisms and their environment, the more useful the belief system, the more it will be used to inform people's actions and views about the universe. Throughout the history of humanity, society's complexity has fed off the information acquired through collective learning. An important difference is that the pace of change that results from collective learning is much faster than the change that occurs in genetic evolution, therefore, human society is highly dynamic and transforming while other species' way of life remains largely the same over time because they lack the superior forms of communication made possible by language and writing. Therefore, they do not display any significant forms of collective learning.

Big History divides the timeline of humanity into three stages which, in order of increasing complexity, are: 1) Paleolithic (250,000-10,000 ya—years ago), 2) Agrarian (10,000-300 ya), and 3) Modern (300 ya-present). The Paleolithic era began when humans first appeared and lasted until about 10,000 years ago, just after the conclusion of the last ice age, and was by far the longest era because it accounted for 96% of humanity.[49(pp99-100)] During this time, people lived in small, kin-based communities and used collective learning to perform their hunter-gatherer duties. The more they learned, the better they were at adapting to environmental changes, and exploring the Earth in search of new environments to settle, which they began doing by about 100,000 years ago—150,000 years after the first humans appeared. During this time, the pace at which collective learning was driving societal change was still slow by today's standards, so aside from adapting to new environments, populations did not make many huge technological leaps during the Paleolithic era. However, the spread of humans meant that other hominid species, such as Ergaster and Neanderthals, were forced to try to survive alongside humans, a task that proved difficult due to the superior mental faculties of humans which gives them a competitive advantage when it comes to extracting the resources that both they and these other hominid species needed for survival. In fact, most experts believe that Ergaster and Neanderthals, as well as many other animal species, ultimately went extinct because of humanity's impact on the biosphere, most notably, its increasing share of biospheric resources.[49(p106)]

The last ice age came to an end approximately 10,000 years ago when glaciers and harsh, cold conditions receded closer towards the poles, allowing once cold regions to exhibit a much milder climate and become suitable locations for populations of humans to settle down and eventually start to combine several techniques, many of them already in existence, to make a new, more powerful technology called agriculture.[98] This newfound ability to extract increased amounts of resources from the environment marked the beginning of the Agrarian era and the universe's ascension beyond the 7th **complexity threshold**. With agriculture, humans began to selectively cultivate plant and animal species that they felt were useful, either for nourishment, or for their ability to do work.[49(pp113-116)] Furthermore, the abundant resources that agriculture produced led to "intensification" — the coming together of larger amounts of people, in close contact, to create dense population centers where the interaction between individuals gives rise to increased collective learning,[49(p119)] and a more complex, hierarchically-structured society reflecting the fact that there were now more ways of life and societal roles, such as a myriad of specialized occupations.[37(p153),49(pp128,138)] The same way gravity created stars out of helium and hydrogen whose fusion in the star's core created the heavier and more complex atomic elements, cities bring people together who then interact to create ever more innovative technologies and novel systems-of-thought.[49(p111),99] By today's standards, these agriculturally-supported population centers would be classified as towns and villages, and as with any system that becomes more complex, society began to display emergent properties that were products of interacting towns and villages, such as trade, warfare, and governmental powers whose job was to coordinate activity within the increasingly complex towns and villages.

Powerful Agrarian societies put pressure on neighboring populations who had yet to adopt agricultural technology as a way of life,[49(p119)] and the same process of agriculturally-based intensification led to the same increase in societal complexity in almost every region on Earth where humans have populated—first Mesopotamia (portions of the Middle East and Northern Africa), then China and India, lastly Europe and the Americas.[49(pp118-119)] By now, the populations in these regions acquired unique physical features which are reflected in the 0.1% variance in human DNA, and furthermore, these regional differences gave rise to cultural differences as well.

By the time humanity reached the Later Agrarian (5,000-300 ya), the intensification occurring in villages and towns had caused many of them to grow into large city states with a more complex societal structure that required even more powerful, knowledgeable, and well-organized governments sitting at the top of the social hierarchy. In addition, humanity started to become more integrated because by now, city states interacted even more through exchanges of plants, animals, people, goods, ideas, etc.—all objects relevant to human life and the functioning of government were starting to be traded between nations, and all levels of nations' societal hierarchy. This led to intensification and collective learning occurring on a much larger scale than it did within individual city states. In short, the nodes of society's large-scale networks were starting to connect, and by the end of the Agrarian era, there were just four distinct world zones — Afro-Eurasian, American, Australian, and Pacific.[49(p108)]

Societal integration is one of the central themes of the next era of humanity, the Modern era (approx. 300 ya-present), whose commencement signaled the universe's crossing of its **8th complexity threshold**. As we've seen, it started out as a gradual process, occurring first on very small scales during the Paleolithic era, shortly after humans left Africa, then expanding and quickening throughout the remainder of both the Paleolithic and Agrarian eras, leading to the four world zones listed above. But more integration would be necessary before virtually all of humanity was interconnected, marking the beginning of an ongoing process of globalization. We will consider the Modern era to have commenced when the four world zones existing at the end of the Later Agrarian became united through exploration, the establishment of trade routes, and other forms of interaction between the established population centers.[49(p185)] This globally integrated human network was made possible by revolutions in travel on land, in the sea, and in the air, which in turn led to revolutions in communication since now goods and information could be transferred across distances that spanned the globe, allowing once separate entities, whether physical or mental constructs, to be combined in new ways to form new systems.[49(pp155-157)] In short, it was now possible for collective learning and intensification to manifest on a global scale.[49(p205)]

From early on, globalization has been facilitated by several forms of technologies that were increasingly becoming more dependent on the insights of science — a newly emerging system-of-thought that epitomizes humanity's desire to understand the universe and is ultimately an outgrowth of one of the most basic of human instincts: to construct conceptual population increases and occur when productivity falls below a level that can sustain the current population. The Modern era was the first time that it was possible to stave off severe widespread Malthusian Cycles, because now, productivity routinely outpaced population demands, and as a consequence, populations increased dramatically during this time.[49(pp189,200-201)] Locally, however, civilizations do rise and then fall, but globally, on the largest scale of society, the overall trend has been an upward climb in population and the level of complexity for the thriving civilizations, behavior that is similar to what's displayed by today's global economy.[49(pp146,152,160,162-163)]

Another key factor that enabled the increase in production was our widespread use of fossil fuels—coal, oil, and natural gas—during the Industrial Revolution, which led to advances in manufacturing, travel, and standard of living. As with every other subthreshold crossed, these advances enabled the global society to experience increased complexification, intensification, and integration. This did not come without a price, however, since the by-product of the work performed by humans through the use of natural resources is large amounts of consumption and a less usable form of energy, consistent with the Second Law of Thermodynamics. Thanks to collective learning, we humans have the ability to transform our environment according to our will, and in fact, are so good at this, we can make it harder for the other species who share the planet, oftentimes pushing them towards extinction. In addition, because of the waste products and other adverse effects of humanity's insatiable consumption of natural resources, we are now contributing to climate change, which has the potential to cause a strong global Malthusian Cycle.[49(pp209-212)]

Nevertheless, today's society, consisting of networked humans and their technology, form the most complex and powerful information processor ever known to exist. The global society now has the ability to send goods and people from one location on Earth to just about any other in only a matter of days, and people can instantly communicate with each other and access information in databases, almost no matter their location on Earth. This all means that there is more extensive societal integration facilitated by faster communication—the nodes of the global societal network share people, animals, products, diseases, ideas, emotions, and events faster than ever before.[37(p153),49(pp175-176),101] This synchronization has sped up the pace of change so that now, fundamental changes to society and ways of life occur on the scale of a single human life span.[49(p204)]

As Above, So Below; As Below, So Above

In this chapter, we have assumed an M-theoretical model of the universe that treats it as a hierarchical system of 10-branes where one brane, the boundary brane, is a hologram whose activity results in the projection of the physical universe through an eleventh dimension to the second brane, the interior. Then we examined the range of complexity displayed by the physical systems constituting the universe over time, from the simplicity of empty space—the vacuum—which only contains the noise-cancelling Higgs ocean and the virtual particles that blink into and out of existence, to the most complex systems of all, humans and the systems that we are elementary units of, such as the global society and economy. We saw that the universe began in the simplest state then evolved such that, in localized regions like the Earth, it was possible for systems that are more and more complex to arise. The right ingredients and conditions existed on Earth so that it became a location within the universe where this complexification was able to cross the threshold of life, and then eventually produce humans and our many civilizations over the course of humanity.

The discussion in this chapter will enable you to see the correspondence between the evolution of complexity in the universe and the processes that occur within the brain, particularly during sleep, that give rise to states of increased complexity.

The appearance of humans made it possible for the universe to consciously model itself, since it is every human's instinct to essentially attempt to model their environment as a means for survival. In addition, our language, reading, and writing have effectively networked us so that a collective knowledge, wisdom, and memory can build up over time, an activity referred to as collective learning. This ability makes us one of the most influential species this planet has ever seen because it enables us to transform our environment according to our needs, making us a very adaptable and resilient species—qualities that we'll need as we, once again, cope with climate change, except this time, we have played a more active role in its occurrence.

In an evolutionary process similar to the one that produced the hierarchy of life forms, our collective learning has produced our collective knowledge and wisdom—a hierarchy of systems-of-thought. Within this hierarchy lies one of the many crown jewels of human intellectual achievement, modern physics, which as it stands, has almost evolved to the point where it can model all fundamental physical phenomena, i.e., we have made great strides towards the mythical "theory-of-everything". To many, M-theory is our best candidate for this theory and it is capable of resolving the differences between our two most cherished theories, quantum mechanics and general relativity, but it contains entities that we have little hope of ever experimentally verifying, such as strings and branes. In fact, the theory is so mathematically rich and complex that without experimental verification, we cannot determine its final form because there are many possible ways that it can be used to construct a perfectly viable universe.

One way to circumvent our inability to obtain experimental observations of the phenomena predicted by M-theory is to find a different kind of model for the universe—a physical one, as opposed to the mathematical one that the final form of M-theory would provide. Physicists have been considering the possibility of super and/or quantum computers being able to model the universe, and they have even considered certain types of crystals that display similar symmetry properties as the universe. With the Brain-Universe Isomorphism Framework (B-U IF), on the other hand, I propose an alternative model of the universe; in Part II of this book, I will use modern neuroscience to define the physical system that I claim is that model—the human brain. After having read through the model of the universe that I presented here in Part I, I urge the reader to take an active role and keep an eye out for parallels between the universe and the human brain when reading Part II.

Part II

Brain

Chapter 8

Transition to the Brain

Form and Function

My principal hypothesis is that the human brain is a **physical** model of the universe which means that the two systems have the same structural organization and dynamics, making the human brain the perfect guide for physicists in search of the theory-of-everything. I believe there is much potential synergy that can be realized between physics and neuroscience if the two communities were to ever open up a constant dialog so that the details of the correspondence could be more thoroughly investigated. To my knowledge, there has yet to be this type of systematic investigation into the existence of structural and dynamical similarities between the human brain and the universe, however, it is already widely believed that one of the human brain's principal **functions** is to form models, or, representations of the environment using sensory perception and information that it has abstracted through experience. This is a fundamental aspect of human nature and occurs both consciously and unconsciously,[7(pp109-110,125,130-131),102,103(p15),104] and the more we interact with our environment, the more we learn and are able to refine our beliefs about the nature of reality — no matter if these beliefs are correct or incorrect. Through experience, each of us is building our own personal model of reality, and because we can communicate using speech and writing, this turns into a collective process — as in collective learning — that allows for the insights of one human to spread to other humans and in-turn, affect their behaviors and thoughts.

As we saw in Part I, one system-of-thought that has arisen out of this collective process is modern science whose ultimate, fundamental goal perhaps is to discover the true nature of reality—a process that relies heavily on the use of conceptual and mathematical models to explain and predict physical phenomena and to check the accuracy of these models by making observations. Similar to how the interactions of each individual with their environment offers an opportunity to learn and update their belief system, in science, differences between prediction and observations are used by theorists to improve their conceptual and mathematical models so that over time, the models more accurately reflect reality.[7(pp182-183)] *Today, in my opinion, modern science is close to the realization that the missing key principle that leads to the long-sought-after theory-of-everything is the fact that the human brain is the model of the universe.*

If we take the ideas emerging out of string/M-theory to represent a convergence towards our most comprehensive and accurate view of the physical universe, does the human brain, or at least portions of it, have the same, or at least similar, structural organization and dynamics as a universe described by string/M-theory? I have embarked along a research trajectory that has allowed me to begin to assess if the human brain is another physical system that can be used to model the universe, alongside the other objects currently considered by physicists like computers and crystals. In this case, the brain would be a particular type of physical model, a living model, one that most-likely can arise within the universe wherever the right conditions exist. If within the human brain a physical model of the universe can be found, then the details of the correspondence between the two systems are capable of illuminating the way forward for theoretical physicists who feel that there is a theory-of-everything out there waiting to be discovered.

If it is true that the human brain and the universe share the same structural organization and dynamics, then it is possible to describe both of these systems using conceptual models and then show a correspondence between the fundamental aspects of the two models. In the first part of this book, I presented a qualitative conceptual model of the universe that is based on some of our most successful theories in modern science and takes into account some of the emerging ideas that show promise for one day being fully incorporated into our understanding of the universe. Now, in the second part of this book, I will present a qualitative conceptual model of the human brain, which is one of the most complex systems ever known to exist and is an elementary unit of systems that are arguably even more complex, such as the global economy and society. *As you read through this part of the book, I urge you to make note of any similarities that you notice between the brain and the universe.* Although I hold off until Part III of this book to explicitly state in detail key correspondences that I see to exist between the two systems, I will in this chapter, however, briefly mention some of the more important correspondences as sufficient detail about the nature of the brain gets introduced. These "waymarkers" will help you navigate through the many details of each model so that you can begin to get a sense of how I've come to see the brain in relation to the universe.

The Triune Brain

We humans exist at the top of life's complexity hierarchy, but this does not mean that we are free from all anatomical subsystems present in less complex species, because as life has evolved, it has had a tendency to build new forms on top of existing forms.[103(pp113-115)] The Triune Brain Theory is based on this phenomenon and it says that the human brain can be partitioned into three interconnected structures that were produced roughly in chronological order during different stages of the evolution of Earth's species.[103(pp280-281)] Over the years, the Triune Brain Theory has been modified due to newly discovered information, and its level of acceptance and popularity within the neuroscientific community has waxed and waned. Regardless, it still offers a convenient and useful approach to categorizing different regions of the human brain.

One of the simplest and most primitive parts of the human brain was produced during relatively early stages of evolution, before the appearance of mammals, and is referred to as the reptilian brain since it is also present in the brains of reptiles; however, it has been found to be common among several other classes of earlier life forms, such as amphibians, birds, and fish, leading many researchers to believe that the appearance of this subsystem can probably be traced back to a common ancestor of all the vertebrates. The reptilian brain is involved in the performance of many of the automatic, yet vital tasks of the body such as breathing, balance, heart rate, etc. The next structure to evolve, the limbic brain, was definitely present in early mammals, but may have also been present in the early vertebrates; it is associated with mental phenomena more complex than those produced by the reptilian brain such as memory-related processes, and emotional processes like fight-or-flight. The newest structure in the brain is the neocortex which appeared in early mammals and is a distinct feature of primates and humans because it is responsible for superior mental functioning such as self-consciousness and rational thought. Its size relative to body mass has increased over the course of evolution, from its first appearance in early mammals, then in the earliest primates, and now currently in humans, but the cell types and connectivity have remained largely the same.[5(p70),103(p50)]

The Triune Brain Theory says that the three structures discussed above appeared sequentially over the course of evolution and have been incorporated, one on top of the other, into a single, interconnected system, essentially giving the human brain a hierarchal structure (Figure 8). For the model presented in B-U IF, I will focus on regions within the upper and middle levels of this hierarchy, namely, the neocortex and areas within the limbic system.

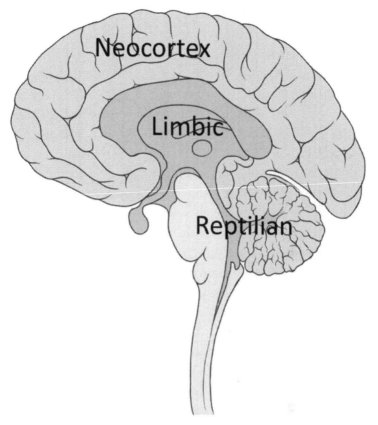

Figure 8

What Informs the Model?

The information used to construct my qualitative model of the neocortex and subsystems within the limbic brain mostly comes from the findings of neuroscience, and more recently, from information theory and nonlinear dynamics as well. When it comes to examining the human brain to determine its structural organization and dynamics, for ethical reasons, most of the work has to be either noninvasive, such as electroencephalography (EEG) or functional magnetic resonance imaging (fMRI), or it must be performed on the brains of deceased individuals. However, researchers have been able to implant sensors inside the brains of people living with certain conditions, such as epilepsy, that require medical intrusions inside the brain. On the other hand, many aspects of my model were obtained by neuroscientists making observations on other primates and mammals because experiments involving these species do not bring about nearly as many moral hurdles as if the same invasive procedures were performed on humans. By systematically acquiring as much data as possible regarding the human brain using the limited available methods for studying humans, and by studying the brains of many other species of animals using a wider array of techniques, researchers have been able to conclude, with strong confidence, many of the phenomena and structures that exist in human brains.[105(pp245,251-252)] Furthermore, most of the information that I will include in this model was obtained by studying the visual system, which has emerged as the subsystem of choice for researchers to focus on because visual information accounts for the largest percentage of the sensory-related processing power of mammalian and primate brains,[5(p14)] especially for the latter whose stereoscopic vision results in a wealth of visual information. The percentage of the brain's real-estate dedicated to processing visual information is an indication of the importance of this type of information for the survival of these species, and fortunately for the researchers attempting to understand the brain, many of the insights learned from studying the visual processing system can be applied to other areas of the brain because it has been found that they share the same general underlying structure and dynamics, and therefore, most likely process information similarly.[5(pp71-72),103(p58)]

Light and the Eye

Recall from the Standard Model that electromagnetic radiation is a manifestation of the electromagnetic force and is produced when the state of a charged system changes. The possible wavelengths of this radiation vary widely and depend on the sources, from the short wavelength gamma rays produced by very energetic processes, such as those occurring inside the nucleus of atoms, to the very long wavelength radio waves produced by lower energy processes. Human vision detects light within a tiny sub-band of the electromagnetic spectrum known as the visible spectrum.

The cells in the eye that detect this light exist within the retina and are called rods and cones. Each rod and cone in the eye has a unique receptive field determined by its sensitivity to particular locations within the environment, the wavelength of light, the intensity of the light, motion of source, etc. Rods are useful in low-light situations because they are sensitive to single photons, and they are involved in the visual representation of the periphery since this is where the light that they detect originates. Furthermore, rods are sensitive to motion but are not involved in the detection of color because they each respond to the same wavelength. On the other hand, there are three varieties of cones that collectively sample three sub-bands of the visible spectrum, providing the information necessary for the visual system within the brain to determine color. The three types of cone cells can be differentiated by the wavelengths that they are sensitive to, i.e., shorter, intermediate, and longer wavelengths of the visible spectrum.[106(pp95-97)] In addition, cones require strong signal levels, or, many photons, and the light that they detect primarily originates in the central regions of the visual scene. All of these rods and cones form functional units in the retina, like an array of pixels that spatially samples the light making it into the eye from the visual scene.[102,106(p50)]

Despite these cells' sensitivities to wavelength, brightness (number of photons), and motion, each provides a representation of the scene that is still not quite ready to be forwarded to the visual system in the brain. Before this can happen, the signals from the rods and cones are sent through an intermediate network of cells, and then to ganglion cells that actually send the signals out of the eye. One of the purposes of this additional processing is to make sure that the information in the three signal components that collectively encode color are disentangled so that the essential information is communicated in signals that are independent of each other, thus, removing all redundancy.[106(pp107-108)] The activity within the complex intermediate network is continuous while the output from the ganglion cells out of the eye and to the brain is sent in incoherent discrete chunks called action potentials that represent a rich amount of information about some particular point within the scene due to the preprocessing performed by the eye.[106(pp54-55)] There is one type of rod cell and three types of cone cells that detect light in the retina, but there are only two parallel visual channels that leave the retina. One channel communicates changes in the contrast and temporal content of visual stimuli but exhibits low spatial resolution and contains little information regarding the color of the light. The other channel contains high spatial resolution and color information but low contrast sensitivity. Upon leaving the retina, these two channels are sent to the thalamus and then on to the neocortex.[107(pp161-162)]

Where the Magic Happens

In this chapter, I have restated the principal hypothesis of B-U IF, that the brain not only constructs models of the universe using the states of billions of networked neurons, it also models it physically as well. After having presented a qualitative model of the universe in the first part of this book, I will do the same for the human brain in this, the second part of the book. In the end, I will present a comparison of the two models showing many parallels. If in fact the human brain has the same structural organization and dynamics as the universe, which I do believe my analysis has revealed, then it could serve as a useful guide for the theoretical physicists searching for the theory-of-everything.

The Triune Brain Theory suggests that the human brain has a hierarchical organization thanks to life's knack for building on top of existing forms over the course of evolution. I have used this hierarchy to define the particular subset of brain regions that are the focus of my considerations, namely, the neocortex and structures closely associated with it in the limbic brain, such as the thalamus and hippocampus. Furthermore, there is a need to clarify the sources of the information that I use to describe the human brain, because due to moral considerations, it involves research performed on other species of mammals and primates. In addition, it is most efficient to focus the analysis on the visual system of the brain because it is the largest and most well-studied sensory subsystem, and insights from studying it can also be applied to other regions of the brain.

The visual processing system within the brain receives signals with high information content thanks to the preprocessing performed by the complex arrangement of rods, cones, intermediate network, and ganglion cells within the eye. The human brain is capable of taking these signals as input and forming explicit representations of objects constituting the visual scene. In the next chapter, we will begin to explore the structural organization of the brain beginning with the neocortex.

Chapter 9

Neocortex

Top of the Triune Brain Hierarchy: the Neocortex

I will not attempt to provide a detailed description of all areas of the brain, rather, I will focus on two of the three subsystems that the Triune Brain Theory says appeared sequentially throughout the course of evolution, namely, the neocortex and two of the subsystems within the limbic brain—the thalamus and hippocampus. I will begin with the neocortex because thanks to its complexity—diversity of cell types, structural organization, and dynamics—it displays most of the characteristics that will be needed to describe the subsystems in the less-complex limbic brain.

The neocortex is the youngest of the brain's subsystems and is a highly convoluted sheet about 3 mm thick that serves as the brain's outermost structure. For the sake of simplicity, its folds can be neglected so that it can be imagined as if it were spread out and laid flat across the surface of a table,[103(p58),108] and to a certain extent, it can be regarded as a relatively homogeneous structure because the variety of neurons throughout it is relatively constant, and computation within local regions of it is fundamentally the same. In addition, the entire neocortex has both a layered and a modular structural organization, a modularity that is multi-scalar because neurons connect to form functional units called minicolumns, minicolumns connect to form columns, columns connect to form hypercolumns, and hypercolumns combine to form cortical areas such as those involved in processing sensory information.[108,109] But upon closer examination, the variance within the structural organization of the neocortex becomes apparent and exists at least in part because the neocortex has been fine-tuned over the course of evolution to extract statistical regularities in the world that are conveyed to it by the body's sensors.[103(p278)] For instance, cell size and density, as well as the amount of intermediate and long-range connections, can vary systematically from area to area, and it is these structural characteristics and circuitry that determine the functional characteristics in each area — i.e., the structural characteristics and circuitry determine the information that is processed by the neurons within any particular region of the neocortex.[103(p177),105(pp370-371)]

It can be useful to view the neocortex as a complex network whose nodes are the cell bodies, or somas, of the many different types of neurons that communicate with each other via links formed by their axons and dendrites.[103(p56),105(pp371-373)] Dendrites are where neurons receive the majority of their inputs from other neurons. The size and branching pattern of a neuron's dendrites determines the number of neurons that it receives input from while the relayed information is dependent on the characteristics of its soma and axon, as well as the states of the targeted neurons.[105(pp371-372)] Communication between two neurons occurs when electrical signals called action potentials are transmitted along axons which exit out the bottom of the soma and extend through the neocortex for distances ranging from very short, as in to an adjacent neuron, to very long, as in to a completely different area of the neocortex or even to the limbic system. The axon can branch considerably near its end to make contact with many different neurons, creating a multitude of termination points, and when an electrical signal reaches one of these end points, it causes a variety of chemical messengers, depending on the particular neurons involved, to be emitted across a tiny gap called a synapse that lies in between the end of the axon and the target neuron. These molecular-based interconnections can change strength and dynamics across many timescales depending on the experiences of the animal,[5(pp10,35)] and communication has officially occurred when the chemical signals trigger activity within the target cell.

In general, neocortical neurons are complex structures with configurations that change over time because they often have many distinct communication channels whose activity—receiving or ignoring messages—is sensitive to particular values of their membrane voltage, or potential. In fact, whether a communication channel is open or not is a probabilistic event determined by this membrane voltage which changes over time depending on the ongoing network activity that the neuron is a part of.[103(pp144-145)] Typically, neurons have a "ground state" corresponding to what's known as a resting voltage; when they are in this state, they do not transmit action potentials to their target neurons. On the other hand, if a neuron integrates enough incoming electrical signals, its membrane voltage may rise above a threshold value, causing it to then fire off action potentials. In a sense, this communication is similar to the kind that occurs within telecommunications technology, namely, the transmission of information encoded in matter and energy that affects the state of systems on the receiving end of the message.[7(pp113-114)] The types of neurons that exist in the neocortex can be classified into two broad categories based on the effect that they have on the membrane potential of their target neurons. Excitatory neurons temporarily bring their target's membrane potential closer to threshold, enhancing the probability that they will fire action potentials while inhibitory neurons do the opposite.[5(p72),103(p62)]

Excitatory Neurons

Pyramidal neurons are one particular type of excitatory neuron that plays an important role in the information processing occurring in the neocortex because they generate nearly all of the excitation initiated there, resulting in the communication of information computed within local circuits to hundreds of neurons both near and far[5(pp74,85),103(p58),105(p367)] (Figure 9). For neocortical pyramidal neurons, action potentials received as input can be either subthreshold or suprathreshold. In the former case, the incoming action potential does not cause the pyramidal neuron on the receiving end to fire action potentials, although it does change its membrane potential by bringing it closer to threshold. On the other hand, in the latter case, the incoming action potential is strong enough to cause the neuron on the receiving end to also fire action potentials that are transmitted to its own target neurons.[109,110] In addition, pyramidal neurons send and receive messages at distinctly different times because their somas and axons cannot integrate incoming action potentials while they themselves are firing action potentials; likewise, they cannot integrate incoming action potentials during a short recovery period after they

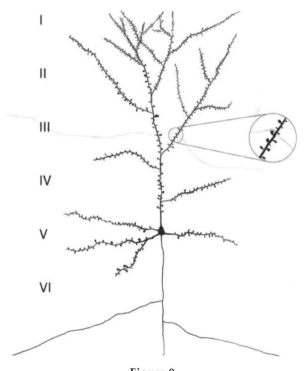

Figure 9

have just fired an action potential. At the conclusion of this recovery period, however, they are once again able to integrate incoming signals until they cross threshold and fire more action potentials. Sustained subthreshold input, which underlies processes like attention, effectively lowers the neuron's firing threshold, making it more susceptible to responding to its preferred input.[5(pp43,157),109] Furthermore, when a neuron fires rhythmically, the time between periods of transmission and reception determines its frequency of transmission.[103(pp139-141)]

Throughout the neocortex, these pyramidal-shaped neurons are oriented vertically such that their prominent apical dendrite ascends out the top of the cell body, or soma (located in layer V in Figure 9), and extends for some distance through the thickness of the neocortex before branching to form an apical tuft (beginning in layer II/III in Figure 9) where input from far and wide is received at voltage-dependent synapses—i.e., the particular termination point of the input to the apical tufts depends on the voltage of the input, or, its strength.[5(pp74,261)] In addition to the apical dendrite, these neurons also possess basal dendrites that can extend out horizontally in all directions from the soma and form a branching pattern to receive input near the same neocortical depth as the soma.[5(p74)] Typically, the dendrites of pyramidal cells are spiny which means that they have tiny protrusions that mark locations where excitatory inputs are received[107(pp7-8)] (enlarged region in Figure 9).

In addition to pyramidal neurons, another type of excitatory neuron that has an important role in the circuitry of the neocortex is the spiny stellate neuron who is just like the pyramidal neuron except it lacks the apical dendrite and it primarily has local connections.[5(p75),103(p58),107(pp7-8)] This type of neuron is involved in the initial processing of much of the subcortical input to the neocortex, as well as input to one neocortical area from another.[5(p72)]

The axons of pyramidal neurons descend downward out

the base of the soma and can project long-range to neurons in distant regions of the brain. Fast action potentials emitted by the soma propagate both forward to the axon collaterals and backward to the dendrites, and once generated, they can travel through the entire axon tree and signal downstream neurons,[103(p87)] but oftentimes not before a copy of the message is given to local neurons via short collaterals that appear in the axon close to the point at which it left the soma.[5(p74)] The speed of action potentials depends on the size of the axon and whether or not it is insulated.[5(p35)] In general, action potentials travel increasingly slower through very long axons connecting distant neurons due to axonal conduction delay.[103(p276)] On the other hand, there isn't much of an axonal conduction delay in the communication occurring between neurons within the same local circuit.[111] Action potentials are a discrete—digital—form of communication as opposed to the continuous—analog—signals associated with membrane and field electrical potentials, and they are the primary way messages are communicated rapidly from one neuron to the next.[5(p35),103(p104)] Because they are more specific than the more diffuse neurochemical communication and local field potentials, action potentials are more suitable for representing the information involved in perception, cognition, and motor action.[5(p36)]

Axonal projections that target the dendrites of other neurons are referred to as axodendritic; those targeting the soma of other neurons are referred to as axosomatic; those targeting the axon of other neurons are referred to as axoaxonal. Some axoaxonal projections give neurons the ability to regulate, or "gate", the transmission of signals between two other neurons.[112] Furthermore, axoaxonal projections can produce what's known as antidromic action potentials, or, action potentials that travel backwards within the axon towards the soma as opposed to away from the soma which is what happens during more typical action potential transmissions. Antidromic action potentials resulting from stimulation of the end point of an axonal projection can make it all the way back into the dendrites of the neuron, allowing these types of action potentials to play a key role in the strengthening of connections between neurons.[113] In the event that two action potentials traveling in opposite directions within the same axon collide, they annihilate each other.[114]

Inhibitory Neurons

There are a variety of inhibitory neurons that collectively play a key role in the timing of pyramidal neuron activity and the computation occurring within the circuits that they form. Some inhibitory neurons only have local connections while others have long-range connections, and the different types of inhibitory neurons target different parts of pyramidal neurons such as the dendrites, or the cell body and initial region of the axon. The latter connections result in the output region of pyramidal neurons being controlled by inhibition, ensuring that the right neurons become active at the right time and that information gets output in the right direction.[5(p76),103(pp65-67,76),107(p61,Ch3)] Furthermore, the state of neocortical inhibition shapes the stimulus-response properties of neurons all throughout the neocortex and is a major reason why the response of the same underlying pyramidal neurons to the same stimulus can be different from one sensory exposure to the next.

The most numerous kind of inhibitory neurons are those that target the soma, providing it its primary form of direct input.[103(pp66,77)] The second most frequent type is the dendrite-targeting ones; the least frequent are ones with long-range projections. In addition, there are inhibitory neurons that only make connections with other inhibitory neurons, some of which do so via a special kind of direct electrical connection called a gap junction that allows for high speed communication, enabling them to play a critical role in the regulation of the timing between excitatory neurons.[5(p36)] Furthermore, inhibitory neurons provide input to pyramidal neurons that is much stronger than pyramid-to-pyramid input; one reason for this is that there are more synaptic terminals connecting inhibitory neurons to pyramidal neurons; another reason is that inhibitory neurons have lower action potential thresholds and fire at a higher frequency than pyramidal neurons.[103(p66)] Typically for pyramidal neurons, the number of inhibitory contacts decreases with distance from the soma while the number of excitatory contacts increases.[103(p67)]

A trend in mammals has been for inhibitory neurons to be distributed unevenly throughout the layers of the neocortex such that their concentration is highest in the layers that receive input from the thalamus.[107(pp65-66)] This places them in an optimum location to participate in the immediate circuitry that receives thalamic input and allows inhibition to have a profound impact on the receptive fields of pyramidal neurons in the neocortex. However, not all inhibitory neurons are part of the immediate circuitry for input into the neocortex because there are others situated in layers outside the termination zones of thalamic input and these inhibitory neurons modulate information flow between the layers of the neocortex.[107(pp96-97)] Consistent with a general level of homogeneity throughout the neocortex, it is believed that similar inhibitory circuits likely exist throughout the neocortex and are involved in all types of sensory perception, as well as higher cognitive functions.[115]

Inhibitory neurons increase the functional complexity of a network of pyramid neurons by changing their properties. For example, they can change the functional characteristics, or synaptic channels, of a neuron by activating and deactivating different regions of it, such as the soma, dendrites, or axons.[103(pp64,67,168)] Furthermore, the different configurations of pyramidal neurons created via inhibition result in different computations being performed. However, it is the proper

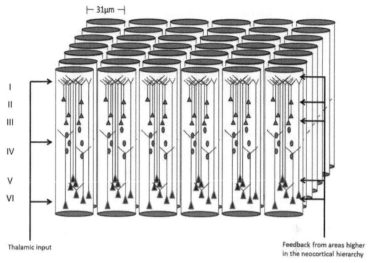

Figure 10

Waymarker #1: In B-U IF, the six neocortical layers within a minicolumn correspond to the six tiny, curled up dimensions in string/M-theory; minicolumn circuitry is analogous to the fluxes that are possible between the six curled up dimensions — the Calabi-Yau shape; the current that oscillates and courses through the minicolumn circuitry corresponds to strings; the several different types of neurons in the minicolumn and the action potentials that they produce are representative of the different types of string vibrations responsible for the particles in the Standard Model of Particle Physics; the "holes" encoded by neocortical circuitry correspond to the "holes" encoded by Calabi-Yau shapes.

Layer 1 is the top-most layer of the neocortex and it receives both thalamic input and feedback from other neocortical areas. The information in these signals may provide the larger context for neurons in layers below whose dendrites extend into layer 1, such as layers 2, 3, 5 and 6 pyramidal neurons. There aren't many neurons whose cell bodies (somas) are in layer 1 and the ones that do exist there are inhibitory;[5(pp71-72),107(p62)] these neurons are the main source of inhibitory input to the apical tufts of pyramidal neurons in layers below.[107(p97)]

Layers 2 and 3 are so closely associated that they are often treated as a single layer (and often referred to simply as layer 2/3). Both are densely populated with pyramidal neurons who have axonal projections that remain within the neocortex,[5(p72)] and apical dendrites that join with the bundles of apical dendrites formed by the pyramidal neurons in layer 5. Therefore, the entire apical dendrite bundle of the minicolumn becomes progressively thicker in layers 3 and 2 before they form their tufts which extend into layer 1.[107(pp11-13)] The pyramidal neurons in layers 2 and 3 are a part of a circuit involving layer 5 pyramidal neurons because their axons descend and give off collaterals (some periodic, or, at regular spatial intervals) that contact pyramidal neurons in layer 5 and they receive reciprocated projections from these same neurons.[107(pp46,51-53)]

Near the border of layers 3 and 4 in many primate species, there exists a network of larger pyramidal and stellate neurons that are in radial alignment with functional units of pyramidal neurons in layer 6.[107(pp18,22,29)] In fact, the axons of these neurons descend towards layer 6 where they travel along the border of layers 5 and 6 and give off collaterals that contact the neurons there.[107(p53)]

Receiving a large portion of thalamic input, layer 4 is considered the input layer of the neocortex and consists primarily of two types of neurons: spiny stellate neurons arranged in vertical strings and inhibitory neurons;[5(p72),107(p5)] neither of these neuron types fit neatly inside minicolumns like the layers 2, 3, and 5 pyramidal neurons do.[107(pp13-15)] In addition to receiving thalamic input into the neocortex, stellate neurons in layer 4 also participate in a circuit that involves pyramidal neurons in layers 2 and 3 and those that are in layer 6.[107(p40)] Stellate neurons are closely associated with pyramidal neurons in layer 6, but in general, they are known to make connections with pyramidal neurons in both the superficial and deeper layers of the neocortex.[107(p73)]

There is an interesting relationship between the size and axonal projections of stellate neurons and their depth within layer 4. The lower the neuron is in layer 4, the smaller it is and the more focused are its projections to neurons in layer 3. Those positioned in the middle of layer 4 are larger and their projections branch more as they ascend towards layer 3. Those positioned at the top of layer 4 are the largest stellate neurons and their projections branch even more to make their contacts near the border of layers 3 and 4.[107(pp37-39,57-58)]

Tall pyramidal neurons exist within layer 5 and are used by the neocortex to communicate to other regions of the brain, such as subcortical targets.[5(p72)] These neurons are central to the minicolumns existing throughout the neocortex and much of the activity there converges onto their apical dendrites,[108] and due to their centrality, their dendrites form clusters that serve as the central axis of the minicolumns.[107(pp11-13)] In addition to their long-range projections, layer 5 pyramidal neurons are well connected with nearby pyramidal neurons because they send projections to and receive projections from layer 5 pyramidal neurons in adjacent minicolumns, and they form connections with pyramidal neurons in layers 2 and 3 within the same minicolumn and those in minicolumns nearby.[107(pp46,51-52)] Similar to the large pyramidal neurons that exist at the border of layers 3 and 4, very large and widely separated pyramidal neurons called inner Meynert neurons — the largest neurons in the neocortex — can exist near the border of layers 5 and 6.[107(pp6,27-28)]

Most layer 6 pyramidal neurons fall in one of two categories depending upon the targets of their axonal projections.[117] Corticothalamic pyramidal neurons in layer 6 send subcortical projections to the thalamus and are in-line radially with the spiny stellate neurons that they communicate with in layer 4 to form functional units, and similar to layer 4 stellate neurons, there is sometimes a relationship between the position of corticothalamic neurons within layer 6 and the destination of their projections—those positioned deep in layer 6 project to the top of layer 4, those in the middle of layer 6 project to the middle of layer 4, and those at the top of layer 6 project to the bottom of layer 4.[107(p50)] The apical dendrites of the layer 6 corticothalamic pyramidal neurons are thicker than most layer 5 pyramidal neuron apical dendrites, and they have apical dendrites that terminate in layer 4 where their lateral spread can be quite extensive.[107(pp12-13,50)] On the other hand, layer 6 also has corticocortical pyramidal neurons, or, pyramidal neurons whose axonal projections stay within the neocortex to target other neocortical neurons. These neurons have widely varying dendrite patterns that largely stay within the deeper neocortical layers—layers 5 and 6. The axonal projections of these neurons extend horizontally and also stay within the deeper layers, often targeting pyramidal neurons in layers 5 and 6.[117] In addition, layer 6 is known to have pyramidal neurons with long and skinny apical dendrites that reach all the way up to layer 1 where they give way to an apical tuft much smaller than the tufts of other neocortical pyramidal neurons. The axonal projections of these neurons target a subcortical structure known as the claustrum and they also extend for great lengths horizontally within the deeper neocortical layers, similar to the axonal projections of layer 6 corticocortical pyramidal neurons.[117]

Waymarker #2: In B-U IF, the lighter quarks of a generation correspond to layer 5 pyramidal neurons and layer 6 claustrum-projecting pyramidal neurons; the heavier quarks correspond to a circuit comprised of any one of these neurons and a layer 6 corticocortical pyramidal neuron within the same minicolumn; electrons correspond to layer 2/3 pyramidal neurons; neutrinos correspond to layer 4 stellate neurons; $W^{\pm}$ and Z^0 bosons

correspond to layer 6 corticothalamic pyramidal neurons;
first, second, and third Standard Model particle generations
correspond to a neuron's depth within its layer—such as
upper, middle, lower; antiparticles correspond to antidromic
stimulation within the axons of neurons; the superpartner
particles in Supersymmetry correspond to neocortical inhibitory
neurons.

Circuitry

The complex internal circuitry of modules in the neocortex
is designed to create new output channels from input
channels and is, for the most part, genetically determined,
although fine-tuning of the connections does occur during the
brain's internal dynamics and through interaction with the
environment.[103(p31),107(p202)] Two types of circuits that exist within
the neocortex-limbic brain structure are the axis (or specific)
and shell (or non-specific) circuits. The axis circuit involves
the layer 5 pyramid neurons that are the central structure of
minicolumns. These neurons form reciprocal connections with
neurons in the thalamus, creating a recurrent or looping circuit
that holds input over time. On the other hand, the shell circuit
involves pyramidal neurons in layers 2 and 3 that surround
layer 5 pyramidal neuron apical dendrites; these neurons form
an input-output circuit via horizontal axonal projections that
connect one minicolumn to another.[108]

One of the ways in which thalamic input is delivered to the
neocortex is via thalamocortical projections that make contact
with layer 4 stellate neurons. From here, the information
is relayed to both the axis and shell circuits. In addition to
this layer 4 input, there is also thalamocortical input to the
neocortex in layers 1 and 6 that also can make it into both
circuits. While the axis and shell circuits are well-identified
and distinct circuits in the neocortex, information does readily
get communicated from one to the other via links such as the

direct contacts between layers 2 and 3 pyramidal neurons with layer 5 pyramidal neurons. In addition, information in the thalamocortical loop involving layer 6 pyramidal neurons is capable of supplying large amounts of activity to the apical dendrites of the layer 5 pyramidal neurons central to the axis circuit—layer 6 corticothalamic pyramidal neurons can project to a thalamic neuron which projects to a layer 4 stellate neuron who then makes contact with the layer 5 pyramidal neuron apical dendrite.[108] In addition, layer 6 corticocortical pyramidal neurons may be involved in the transmission of layer 6 thalamic input to layer 5 pyramidal neurons since their axons have ample opportunity to make contact with the basal dendrites of layer 5 pyramidal neurons.

In addition to the layered and modular organization of the neocortex, it is also hierarchical in that posterior areas are concerned with largely subconscious primary sensory information, intermediate areas are responsible for the explicit representations underlying our percepts, and frontal areas deal mainly with executive functions that are also largely subconscious. Furthermore, pyramidal neurons become more complex the further up the processing hierarchy you go, meaning, they receive more excitatory input, have a more extensive dendritic branching pattern, and have a larger axonal patch size.[105(Ch15)] Information sent up the hierarchy—from the back of the neocortex towards the front—is considered bottom-up, or feedforward, while information sent down the hierarchy—from the front towards the rear—is top-down, or feedback.

The exact nature of thalamocortical input to the neocortex depends on whether it occurs in primary sensory areas on the bottom of the neocortical hierarchy, or towards the intermediate and upper levels because going from primary to higher sensory areas, there is a decrease in the number of layer 4 stellate neurons in the neocortex and an increase in synaptic connections resulting from thalamocortical projections to distal regions of layers 2, 3, and 5 pyramidal neuron apical dendrites, suggesting a shift from the importance of the axis circuit to the importance of the shell circuit.[108] This is parallel to the shift away from bottom-up influence to top-down influence as we move from primary sensory areas to intermediate ones. The implication is that the many stellate neurons in layer 4 of primary areas are needed to process the incoming sensory information, but once this has occurred, and information is communicated higher into the neocortical hierarchy, the importance of processing incoming sensory information is reduced and top-down information becomes more important. Ultimately, however, there are two major kinds of input to the minicolumns in higher sensory areas, the ones that are responsible for explicitly representing our percepts: bottom-up sensory information that ascends the neocortical hierarchy and top-down internal information that descends the hierarchy communicated by the frontal areas of the neocortex.[108]

There are also lateral connections, such as the periodic ones occurring in layers 2, 3, and 5, that couple neurons and minicolumns encoding for similar information at the same level in the neocortical hierarchy, allowing for information to spread and for larger structures such as columns and hypercolumns to form.[5(p120),105(pp163,183),107(pp127,226-228),109] A neocortical column is a cluster of approximately 100 minicolumns connected by these horizontal connections[108,118] and hypercolumns are the clustering of several columns that are collectively capable of representing the many aspects of our individual percepts.[107(pp144,346),109] Lateral connections can originate in all layers that project out of a neocortical area, that is, from all layers except 1 and 4, and can terminate throughout the depth of the neocortex in the recipient areas.[5(p120),105(p183)] Furthermore, these connections not only facilitate the spread of information via excitation, they also contact inhibitory neurons to suppress the activity of groups of competing neurons, enabling the most dominant neuronal coalitions to function autonomously.[103(p63)]

The length of axonal projections in the neocortex varies from very short local connections between neighboring neurons to long distance ones connecting widely separated regions of the neocortex. The number of connections decreases exponentially from the shortest to the longest, meaning that there are many more local connections than long distance ones, consistent with there being localized dense connectivity between the neurons constituting minicolumns and columns but far fewer, yet, precisely directed long-range axonal projections connecting widely separated regions of the neocortex.[103(pp51,54)] Many of these connections are reciprocated so that there are multiple parallel loops of varying length and temporal scales within the neocortex and between regions within it and other closely associated structures such as the thalamus and hippocampus. Therefore, the amount of synaptic and axonal conduction delays present within the brain's communication varies considerably.[103(pp134,276)] This "scale-free" reciprocal connectivity in the neocortex can be described as a fractal of loops.[103(pp30-32)]

In order for separate entities to combine and form a single system, they must be able to interact, or, communicate in some way. The different levels of the neocortical hierarchy — posterior, intermediate, and frontal — are able to form one combined system because they are linked by feedforward and feedback connections[5(p124),103(p164)] that differ significantly from each other and come in several different forms (Figure 11). The usually uninsulated and small caliber lateral projections are sufficient for local circuits but the larger caliber and insulated feedforward and feedback axons are needed to form the long-range connections between different levels of the neocortical hierarchy. Consequently, feedforward and feedback connections are used to transmit messages between the levels of the neocortical hierarchy and they do so at speeds that are at least ten times faster than lateral connections.[105(pp196-197)]

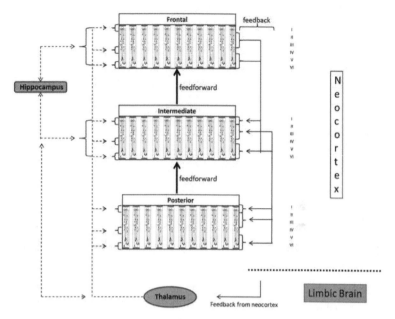

Figure 11

Layer 2, 3, and 5 pyramidal neurons on the same level of the neocortical hierarchy send information to higher levels through feedforward projections that terminate mainly in layer 4, the input layer of the neocortex. In addition, layer 5 pyramidal neurons can send information to higher areas via the axis circuit which means that thalamic relay neurons are involved.[108] In general, feedforward signals are involved in the processing of sensory-related information such as those that get sent from the thalamus to the primary visual system in the back of the brain, then from there to the secondary and intermediate visual centers, and then on to the executive areas in the front of the brain. Feedforward projections are considered driving, meaning that they tend to cause their target neurons to fire in response, a property that is a product of their focused nature and high axon terminal (or bouton) density.[5(p126),105(pp186-188)]

On the other hand, feedback axons are known to originate from pyramidal neurons in layers 2, 3, 5, and 6 and target the same layers in lower levels of the neocortical hierarchy.[105(p388),107(p240)] In fact, feedback is received by all neocortical layers except for the main input layer—layer 4—and the nature of the feedback varies depending on the layer that is targeted. Feedback to layer 1 is widespread and non-specific because it acts on large populations of neurons while feedback to other layers, such as layers 2, 3, and 5, can be reciprocal, meaning that it can match the precision of feedforward projections.[105(p195)] Feedback terminating in layer 1 has a modulatory effect on the response of its targets—pyramidal neurons in deeper layers—as opposed to a driving one, mainly because it terminates on the distal parts of the dendritic tree. Nonetheless, the voltage-dependent currents found in the tufts of pyramidal neuron apical dendrites makes these neurons highly susceptible to coincident feedback from neurons on higher levels of the neocortical hierarchy.[5(p261)] Feedback that terminates on the superficial layers of the neocortex is capable of spreading information laterally within a level of the neocortical hierarchy to a greater extent than the local lateral projections.[105(p193)] Furthermore, the effectiveness of the local lateral projections can be switched on or off by feedback from higher areas, oftentimes indicative of awareness of a particular task.[105(p164)] Unlike the widespread feedback to layer 1, feedback that directly targets deeper layers can be very precise and strong, reliably driving their targeted neurons similar to feedforward projections who have higher bouton density and less divergence than their feedback counterparts.[105(pp186-188,195)]

Typically, information already present in the neocortex, such as memory traces, are relayed through feedback connections from higher levels of the neocortical hierarchy such as the prefrontal cortex.[103(pp164,370),107(p458)] The feedback that terminates in layer 1 provides the larger context for the targeted neurons and feedback terminating below layer 1 produces a more driving influence that areas higher in the neocortical hierarchy have over lower ones during processes like memory.[5(pp71-72),105(p188)] In the absence of sensory input, or even when the sensory input is fragmented, such excitatory feedback is required to maintain specific activity and to fill in missing information within the neocortex, enabling prior knowledge to be incorporated into the present. For example, because of feedback from the higher levels of the neocortical hierarchy, the response in the areas of the brain that represent the information underlying our conscious perceptions is just as selective and almost as strong during imagined imagery as it is when we actually view a scene. However, this feedback does not directly reach the lower levels of the visual neocortical hierarchy, such as the primary or secondary visual cortex, therefore, the lack of activity in these regions results in less spatial selectivity, and hence, imagined imagery that is less vivid.[5(pp300-301)]

Waymarker #3: In B-U IF, the levels of the neocortical hierarchy—posterior, intermediate, and frontal—correspond to the membranes, or branes that exist in M-theory; the feedforward and feedback projections that connect the levels of the neocortical hierarchy are analogous to the 11th extra dimension added by M-theory that allows branes to interact and form systems of branes.

Information Processing

In this chapter, we have explored the structural organization of the neocortex by considering the different types of neurons that exist within it and the ways in which they are organized into functional units and communicate with each other. Both excitatory and inhibitory neurons are arranged either in or around functional units called minicolumns and are distributed systematically throughout the six layers spanning the neocortical depth. Pyramidal neurons are the principal excitatory information processor while stellate neurons are excitatory neurons that play a large role in the initial processing of input to neocortical areas. Inhibitory neurons also play a role in the processing of neocortical input but they do so in a way fundamentally different from excitatory neurons—unlike excitatory neurons, inhibitory neurons reduce the chances that a target neuron will fire an action potential. This is a quality that is necessary to balance excitation in the neocortex and to assure the proper timing between cooperating neuronal modules. Furthermore, inhibition is needed to segregate competing neuronal coalitions and establish dominance in a winner-take-all fashion.

We also saw in this chapter that there are multiple types of hierarchies that exist in the brain. Along with the Triune Brain hierarchy, there is one that exists solely within the neocortex. For example, in the visual processing system, areas in the back of the brain—its posterior—are largely concerned with the processing of initial visual input that we are not directly conscious of. As the information is sent higher in the neocortical hierarchy, it reaches intermediate levels where explicit representations of our percepts are formed; the activity in this level is believed to underlie our direct conscious perceptions. The highest level of the neocortical hierarchy contains frontal areas of the neocortex. Here, executive functions and short-term memory processes take place. All of the levels of the brain's hierarchies are connected by feedforward and feedback projections, enabling them to come together and function as a single system.

This chapter allows you to begin to see similarities between the structural characteristics of an M-theoretical universe and the structural characteristics of the brain.

Next, we turn our attention to a more detailed look at how the neurons in the brain behave by themselves and as part of coalitions of neurons. Doing so reveals important insights about the way the brain processes information.

Chapter 10

Neocortical Dynamics

Coalitions of Neurons

In this very local region of the universe that we call Earth, universal evolution has led to life forms that are capable of using various forms of sensation to extract environmental information that is then used by their brains to form neuronal representations of the environment and events that have occurred, as well as plan for future actions. Furthermore, the brains of Earth's species are some of the most complex physical systems ever identified and are capable of manifesting an innumerable amount of distinct states over the lifetime of an organism. These states are defined by factors like the strength of synaptic connectivity between neurons and the particular combination of neurons and/or functional areas that are in communication with each other. Genes determine the initial structural organization of brains and the intrinsic connections between neurons, which start out weak but are continuously modified based on experience—the least active connections are weakened, or even eliminated altogether, while those that are repeatedly activated are strengthened.[5(pp168,309),103(pp246-247)] In general, connections between neurons are strengthened whenever they fire together, such as when one neuron receives synchronous action potentials from other neurons that lead to its own discharge. Through this process, coalitions form and are fortified, making it more likely that the neurons within them will fire together the next time their preferred stimulus is presented, or the next time the spontaneous activity within the network they are embedded in is favorable for their activation.[103(p247)] Thanks to this experience-based strengthening of connections, over time, spontaneous network dynamics begin to reflect prior sensory and cognitive representations, allowing for the development of neocortical attractor states that are manifested by coalitions.[103(pp164,221)]

A neocortical coalition consists of many cooperating systems of neurons, or, neuronal assemblies, such as minicolumns and columns, whose activity is coordinated in such a way that they perform a specific function, like the representation of sensory or cognitive information. Individual neurons are strongly influenced by the state of the assembly that they are embedded in so that if the assembly is in a state that is favorable for inducing the neuron to fire, the probability of it doing so will be high, which is another way of saying that neurons become enslaved to the larger network-level activity of the assemblies that they belong to. This isn't the end of the story, however, because the assemblies themselves are under the control of even larger systems of interconnected neurons, such as entire neocortical areas, or coalitions that can consist of individual assemblies linked by long-range axonal projections and spread throughout the different levels of the neocortical hierarchy[103(pp268-269)] (Figure 12). In short, on all spatial scales of the brain, systems of neurons—a single neuron, minicolumns, columns, hypercolumns, etc.—are embedded within larger systems of neurons that they can influence and also be influenced by.

Neurons within well-connected populations, such as minicolumns and columns, often display coordinated activity because they have similar receptive fields. However, neurons need not be directly connected to each other to display such coordinated activity because they can be brought together by other means, such as common inputs or the influence of inhibition. In fact, neocortical assemblies can form coalitions through synchronous activation without possessing direct links between each other; coalitions that form in this way are more flexible than the hardwired kind and can form dynamically

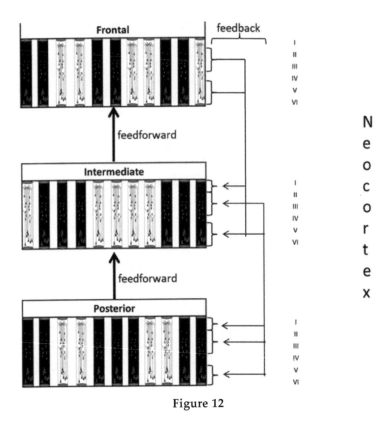

Figure 12

to deal with novel input. Furthermore, they can still have the same influence on target neurons as coalitions that are directly wired together, because to a target neuron, all that matters is that it receives incoming signals within a given critical time period sufficient enough for its membrane potential to cross threshold—it does not care whether the neurons targeting it are connected or not.[103(pp163-164,238)]

Coalitions of neurons carry out the large-scale information processing that occurs in the brain. Any one percept, real or imagined, corresponds to a coalition distributed throughout the neocortical hierarchy, and in order for us to become conscious of this percept, its supporting coalition usually has to win a competition for dominance with coalitions encoding for other percepts vying for attention. During this competition, coalitions reinforce and synchronize the firing activity of their member neurons, and they send signals to inhibitory neurons to suppress their competitors. In this way, winner-take-all phenomena play a key role in the dynamics of coalitions. Furthermore, the portions of coalitions that exist on the higher levels of the neocortical hierarchy, such as those in the prefrontal cortex or anterior cingulate, can exert the influences mentioned above on the activity of coalition members on lower levels of the neocortical hierarchy via feedback.[5(pp305-306)]

Waymarker #4: In B-U IF, coalitions of neurons, whose member assemblies are spread throughout the levels of the neocortical hierarchy, correspond to coalitions of strings that are spread throughout the different levels of the Holographic Universe — the boundary and interior branes. Coalitions of neurons underlie our percepts like coalitions of strings underlie physical matter.

Oscillatory Communication

Excitation and inhibition in the brain provide the positive and negative forces, respectively, that balance each other and make it possible for the manifestation of self-organized oscillatory rhythms of action potentials — the fundamental basis of communication between regions of the brain during processes involved in sensory, memory, or other cognitive representations.[103(pp136,335)] The neocortex is in many ways a complex network comprised of a variety of neurons who communicate via action potentials transmitted along axons and received by dendrites and somas. When the neurons comprising this network are considered alone, they are found to have their own intrinsic frequencies at which they like to send and receive signals, but when their participation in larger populations of neurons is considered, the frequency and timing of each neuron's oscillations is largely determined by the global rhythm that arises from the activity of all neurons in the network. In other words, all participating pyramidal and inhibitory neurons combine to form a single oscillator and no single neuron can be considered as the leader of the rhythm. More specifically, the oscillations of the individual neurons in the network become enslaved to the global rhythm so that they fire phase-locked to it.[103(pp107-108)] This type of enslavement is made possible because interconnected neurons can communicate with each other through both sub- and supra-threshold action potentials and influence the timing of each other's oscillations. The global rhythm persists because neurons within a network exchange timing signals when the phase relationship between their oscillatory patterns is unstable—if one neuron's pace slows, the signals from the other neurons speed it back up and vice versa. On the other hand, as this phase relationship becomes more stable and there is a high degree of coherence within the population, the need for the neurons to exert this type of push or pull on other neurons can decrease.

Discrete Processes

Ultimately, the communication underlying sensory and cognitive information processing within the brain occurs through the transmission of action potentials, which is a discrete process. Furthermore, systems of neurons in the brain process incoming signals in discrete temporal windows determined by the ongoing global network oscillation. All signals received during the same window, or, processing period, are considered synchronous and are integrated, or bound together, while those that arrive in different periods are not.[103(pp150-151),119,120] Ultimately, the purpose of a group of neurons emitting action potentials so that they arrive at the same target synchronously is to transmit messages downstream in the most effective way possible.[5(pp42-43,46,179,309),103(pp137,246)] However, there are limitations to the extent of synchronization due to the finite amount of time it takes neurons to communicate and also because of axonal conduction delays that accumulate for signals that travel far. Therefore, the oscillation frequency of a population of neurons determines the size of the assembly that they can form such that fast rhythms result in small, local assemblies because less neurons can communicate with each other during the short processing period, while slower rhythms allow for larger, more global assemblies because now more neurons can communicate with each other during the longer processing period.[103(pp115-117,122)]

In general, neurons can be brought together into these short temporal windows by two mechanisms that both exert the same effect on their targets. One way is through strong input from the physical world which resets the phase of ongoing oscillations, producing synchrony among the responding neurons. In addition to this type of synchronization, emergent, self-generated synchrony can arise, an example being the case when feedback is applied from higher levels of the neocortical hierarchy to levels below, inducing synchronous oscillations among coalition members and increasing the chances that the coalition will win out and have the information that it represents make it into consciousness. This is especially the case when more than one input vies for attention,[5(p48),110] however, this synchrony is required only initially because it has been observed that when one coalition wins the competition with its competitors who encode for other stimuli, synchrony is no longer required for it to remain dominant, at least until another more salient stimulus is introduced.[5(p46)]

Information is bound together in the brain in at least three different ways. Epigenetic binding is made possible through information stored in our genes that essentially provides the blueprints for the structural organization and intrinsic connectivity of our brains; the information underlying this form of binding has been assembled by life throughout the course of biological evolution on this planet. Alternatively, overlearned binding is accomplished when connections are formed and strengthened by the experiences that we have throughout the course of our lives, creating coalitions well-suited to represent common features of our environment.[5(p169)] The third type of binding is needed because there will always be situations in which we encounter novel stimuli. Synchrony between neuronal assemblies distributed throughout the neocortex appears to be a viable method to flexibly bind novel stimuli since the neurons actively involved in representing this type of stimuli, those whose exact combination of features are new to us, are unlikely to be strongly connected together. Furthermore, it has been observed that spatially distributed neuronal assemblies without direct connections, but with overlapping receptive fields and responding to the same object, display a high degree of coherence[5(pp43-44),103(pp240-241)] and exert the same influence on their downstream targets as if they were directly connected. The binding of novel stimuli probably needs focal attention which is mediated by synchrony-inducing feedback produced by coalition members positioned in the upper levels of the neocortical hierarchy.

Waymarker #5: In B-U IF, the oscillatory activity of neurons, which are based on rhythmic membrane potential oscillations and transmissions of action potentials, corresponds to string vibrations.

Resonance and Filtering

Each neuron has its own inherent natural frequency that it likes to oscillate at but when neurons combine to form larger structures, they produce a global rhythm that they all get enslaved to. For both an individual member neuron and an entire assembly, input signals can be provided at the preferred frequency and phase to cause vigorous oscillation. The closer the frequency and phase of the input signal to the natural frequency and phase of the neuron or assembly, the more successful it will be at inducing vigorous oscillation. This is known as resonance. On the other hand, the larger the difference between the frequency and phase of the input, the less the receiving neuron or assembly will respond. For this reason, neurons and the assemblies that they form can act as filters to ambient signals, selecting only the ones at their preferred frequency to respond to. Pyramidal neurons in the neocortex are capable of performing several types of filtering functions that are a result of the voltage-gated ionic conductances present in their dendrites, although, it has also been found that nearly every part of a neuron can function as an oscillator with resonant properties. If unchecked, resonant activity can lead to runaway oscillation; thankfully, however, one of the main functions of the inhibitory networks is to prevent such runaway behavior. The two main types of neocortical neurons—inhibitory and excitatory—have different resonant properties that complement each other, namely, inhibitory neurons respond best to fast input, while pyramidal neurons prefer slower rhythmic input.[103(pp142-143,145-148)]

It should also be mentioned that researchers have been modeling neurons in the visual cortex as filters that are sensitive to a small range of image attributes such as motion, orientations, and spatiotemporal frequencies. In this view, the amount of action potentials produced by neurons is proportional to the strength, or prominence, of the image attribute that the neuron responds to. Modeling neurons in this way has been successful at reproducing many of the phenomena displayed by the visual system, and in fact, the functioning of whole assemblies may serve as filters, such as a matched filter where an internal representation of an image is used to inspect the incoming sensory information for a match.[5(p177),105(p314),106(p58),121]

Brainwaves

Oscillations exist in the brain at all frequencies from 0.02 Hz to 600 Hz and are generally classified into bands that are each generated in a variety of ways and are associated with different aspects of the brain's information processing.[110] The most commonly identified bands are roughly defined to be delta (1-4 Hz), theta (4-7 Hz), alpha (7-14 Hz), beta (14-30 Hz), and gamma (30-100 Hz), and various types of these oscillations, transient or sustained, can emerge from the same or from different neuronal substrates simultaneously, allowing for assemblies and coalitions of all sizes to coexist and interact.[103(p351)] This is because the tempo of oscillation determines the size of the assembly involved such that fast rhythms result in small assemblies because less neurons can communicate with each other in the short window of time, while slower rhythms allow for larger assemblies because more neurons can communicate with each other in the longer windows of time.

Both slow and fast waves are common in the visual pathway which is comprised of portions of the neocortex and a structure closely associated with it, the thalamus. In this system, alpha waves are the slow waves and, therefore, are more global and travel throughout the neocortex with well-defined phase lags between distinct areas. The populations of neurons involved in the visual system are prone to displaying coherent alpha oscillations when there is a lack of environmental inputs, most likely reflecting internal mental operations such as imagery and free associations.[103(pp340-341)] When environmental input returns, the coherent alpha activity gives way to an increase in localized fast gamma activity which is involved in the processing of the resultant sensory information. In fact, this type of alpha desynchronization is associated not only with sensory perception but also with motor, memory, and other cognitive functions.[5(p39),103(p131)] There are differences, however, in the characteristic oscillations for perceptual and memory systems, namely, alpha oscillations are associated with thalamic activity and the visual perception system while theta oscillations are associated with the activity of another structure closely associated with the neocortex, one that supports memory and navigation—the hippocampus.[5(p39),119] The firing patterns of neurons in the neocortex and its closely associated structures are often organized according to these alpha and theta waves.[103(p267)]

Scientists now realize that there isn't much random communication, also known as random noise, occurring in the brain and that not only a neuron's oscillation rate but also the exact timing of its action potentials encode information.[103(pp157,259)] The existence of the broad spectrum of neuronal oscillation frequencies makes it possible for coalitions of neurons to encode information in many ways. The types of coding that take place in the brain range from sparse coding where one or a few neurons respond to a stimulus, to population coding where large groups of neurons collectively encode information. In population coding, individual neurons do not have to fire the same number of action potentials, respond at the same time, or even contribute a lot of action potentials to the collective activity of the group—sometimes all that is required is just one or two action potentials that are precisely tuned to the proper phase of the collective oscillation.[5(pp30-31,46),103(p239)] One impressive example of this type of coding is multiplexing of signals such as consecutive gamma cycles (higher frequency signals) nested inside alpha or theta cycles (lower frequency signals). This type of information transfer allows for multiple messages to be transmitted simultaneously on the same channel. When the carrier is an alpha wave, which is often the case with sensory perception, roughly 5 gamma cycles can be multiplexed. This is in good agreement with the number of objects and object attributes that a person can simultaneously perceive. On the other hand, when the carrier is theta, which is often the case in memory-related processes, roughly 7 gamma cycles can be multiplexed. This type of multiplexing can support episodic memory, short-term memory, and comparison of current sensory stimuli with a memory trace. Furthermore, the number of gamma cycles that global population rhythm. This mechanism allows the brain to bind information coming from different sources into a single percept.[103(pp10,156),110,119] Systems of neurons have adjustable thresholds and integration periods during which incoming signals are bound together, and the extent of this variability depends on the population rhythm and features of the stimulus such as its intensity and saliency.[5(Ch15)] Since these discrete data chunks underlie our conscious perception and their duration can vary, it should come as no surprise that our perception of the passage of time can vary as well. In fact, our perception of the passage of time may be related to the rate at which these discrete data packets occur. Say for instance that typically five data packets can occur in 1 second; this means that five data packets typically correspond to 1 second of perceived elapsed time. We would perceive time to speed up if the duration of these discrete packets increased so that there would be less of them that could fit within a 1 second period. On the other hand, if the discrete packets became shorter, more of them would fit within the 1 second interval and we would perceive time to elapse at a slower rate—something often reported by individuals who have survived traumatic, near death events.[5(p267)]

Typical processing periods can vary between 20 and 200 ms[5(pp264-265)] and the qualities of a visual scene that are bound together are constant in one of these integration periods, such as brightness, color, depth, motion, etc.[110] For sensory perception, the organizing global rhythm is the alpha band, i.e., the information underlying sensory perception is chunked into discrete "snapshots" determined by the alpha rhythm. Furthermore, external sensory input resets the phase of the alpha rhythm, marking the beginning of a new integration period.[5(p266),119] And counterintuitively, motion is represented by neurons, most likely via their spiking rate, during one individual processing period, or snapshot, as opposed to the difference between consecutive snapshots as it is with the frames comprising a movie.[5(pp264-265)]

Waymarker #6: In B-U IF, the global rhythm of a coalition of neurons defines a clock, one whose ticks—rate of elapsed time—is variable because the global rhythm itself is variable; analogously, the activity of a coalition of strings encoding for matter creates an internal clock, the same clock of Special Relativity, whose rate of elapsed time is variable.

The Edge of Chaos

The oscillatory patterns displayed by neocortical coalitions vary from simple to complex. The simple patterns are very predictable, such as when all neurons in the coalition fire synchronously; the complex patterns are much less predictable since they are characterized by multiple frequencies and phases present within the underlying neuronal population. In fact, complexity and synchrony compete with each other within neocortical coalitions so that when synchrony increases, complexity decreases and vice versa.[103(pp55,164)] Furthermore, it has been found that the oscillatory characteristics of small portions of the neocortex can be the same as those of the entire neocortex—small patches can have the same relative power distribution within the brainwave spectrum as the entire neocortex. In this sense, neocortical dynamics lack a characteristic scale, meaning that similar fluctuations exist on multiple spatiotemporal scales within the neocortex—it is both scale-free and self-similar. This gives the neocortex fractal characteristics and it is consistent with the scale-free and fractal nature of neuronal connectivity within the neocortex. In fact, the scale-free oscillatory dynamics is in many ways a product of the scale-free connectivity. Therefore, neocortical populations that are subsets of the entire neocortex can produce similar patterns of activity as the neocortex as a whole.[103(pp126-127,134,252,276)]

There are many different oscillators in the neocortex—

localized regions of dendrites, individual neurons, whole assemblies of neurons, etc.—that can collectively generate all frequencies from 0.02 Hz to 600 Hz.[103(p113)] Within this spectrum, discrete oscillation bands can be defined that correspond well to, and expand upon, the more conventional brainwave bands mentioned above. Furthermore, the mean frequencies of the brainwave bands are consistent with the scale-free nature of neocortical oscillations because consecutive bands have mean frequencies that maintain a constant ratio of 2.17 which is an irrational number, causing the states produced by global neocortical dynamics to be nonrepeating and quasi-periodic. This is viewed as a weakly chaotic state and lies right in the middle of highly disordered/complex and highly ordered/simple states, giving the brain the ability to efficiently switch between the two extremes.[103(Ch5)] This is an important characteristic since complex and chaotic neocortical dynamics are important for some types of mental processes, while the highly coherent and predictable states are important for others.

It is important to note that the meaning of the word chaos as used here, and elsewhere in this book, differs from the more common meaning of the term that we hear in everyday conversations which typically refers to complete disorder or randomness. The technical meaning of chaos refers to a very unique state of a deterministic, complex, and dynamical system, a state that is neither totally predictable nor totally random. Deterministic chaos represents the "golden mean" between these two extremes and is a characteristic of the most complex state that natural systems can manifest. For the neocortex, chaotic dynamics enable flexible responses, even for uncertain situations, thanks to the multitude of attractor states that such dynamics can support.[103(Ch5)] A system manifesting chaotic behavior can seem predictable to us in the short-term but our inability to perfectly characterize the initial state of the system introduces uncertainties that, over long periods of time, eliminates our ability to accurately predict the state of the system, making it appear as though the system is behaving randomly. It is important to keep in mind that despite the seemingly unpredictable behavior that can be associated with chaotic systems, order is still present because these systems are fundamentally deterministic. Thanks to this feature of complex systems, order can sometimes reveal itself through the manifestation of well-defined global attractor states even after chaos has set in; such states are often referred to as "chaotic attractors".[26(pp38-39)]

Spontaneous vs. Evoked Activity

Neurons can be synchronized, or, brought together into short temporal windows by two general mechanisms that both have the ability to reset the phase of targeted oscillators. The first is strong sensory input. The second is emergent, self-generated synchrony which can be mediated by feedback originating from higher levels of the neocortical hierarchy to induce the synchronous oscillations that help a coalition achieve dominance over competitors.[103(pp254-255)]

In the absence of environmental stimuli, neocortical networks wander non-randomly through the dynamically-generated neocortical attractor states defined by the particular combination of neurons that are actively communicating via sub- or supra-threshold signals. These states, however, do not represent random neocortical noise because they often resemble the states that occur in response to environmental stimuli, and in fact, these internally-generated states have an effect on our ability to perceive environmental stimuli—the closer the match between the internally-generated states and the states induced by environmental input, the better our chances of perceiving the stimulus. This is a form of amplification of the incoming signal, or, resonance between internally-generated states and those induced by external stimuli. If the internally-generated states are far from the states that are evoked by environmental input, then the input could get ignored if it is not strong enough to reset the ongoing oscillation.[103(pp269-271),109,119,122] For weak stimuli at the edge of perceptibility, these internally-generated states act as a gate, or filter, to the incoming sensory signals. If the state of the neocortex is in a permissible state, such as the right attractor or the right combination of oscillations like alpha and theta, then its response to the weak sensory stimuli will be large enough to evoke conscious perception. Experiments have been conducted that reveal that not only is it necessary for the ongoing dynamics to be permissive to the just-barely-perceivable stimuli, or, near-threshold stimuli, the presentation of that stimuli must also occur on a particular phase of the ongoing oscillation. When stimuli are presented on the proper phase of the ongoing oscillations, it induces the phase-locking necessary for the stimuli to be perceived. This phase-locking makes it possible for the information to make it from the first stages of sensory processing, where it is subconscious, to the higher stages underlying conscious perception.[103(p266),105(p192)]

The full range exists from strong and salient stimuli that can completely change the ongoing dynamics of the neocortex, to near-threshold stimuli that do not alter the ongoing dynamics much except for possibly resetting its phase.[103(pp267-268)] At any rate, thanks to the ongoing, internally-generated rhythms of the neocortex, its response to a stimulus can vary considerably from trial to trial.[109] These ongoing dynamics most likely relate to attention, behavioral and consciousness states, memory retrieval, and other aspects of conscious or subconscious cognitive function.[5(p269),109] Furthermore, the prevalence of any particular attractor state depends on its relevance to the task at hand and to its history, or, how many times and how recently it has become manifest, whether through conscious thought, or exposure to the appropriate stimuli. Each time a state manifests, the chances of a similar state making an appearance during the time varies as well.

The neocortex is self-similar and scale-free which means that small portions of it can display the same oscillatory characteristics as the whole. Furthermore, the neocortex appears to be perched in a state perfectly between complex and simple dynamics, which allows it to switch between the two depending on the type of information processing required, such as the processing of internal vs. external sensory information. The neocortex often spends time cycling through attractor states that can resemble responses to external stimuli, and in fact, these attractor states profoundly affect the response of the neocortex when external stimuli are presented, consequently, impacting what we ultimately perceive.

This chapter prepares you to see the correspondence between the "clocks" that constitute neuronal coalitions and the "clocks" attributed to physical systems by Einstein's special relativity. In addition, you've been shown that the brain has self-similar qualities like what is necessary for the universe to have if the central hypothesis of B-U IF is true, namely that the human brain is a model of the universe.

Thus far in Part II of this book, we have focused on the neocortex. We will now turn our attention to two areas within the limbic brain that are closely associated with the neocortex: the thalamus and the hippocampus. By doing so, we will gain further insight regarding the networking of neocortical areas and the brain's ability to store and recall information.

Chapter 11

Closely Associated Structures

Middle of the Triune Brain Hierarchy

The neocortex is the newest structure in the brain and is responsible for the most advanced mental capabilities displayed by some of Earth's species, such as reasoning, consciousness, and self-awareness. It is able to support these complex mental operations thanks to the complexity of its structural organization and dynamics; however, it does not create this magic all on its own because it works very closely with older areas within the brain, such as the limbic and reptilian subsystems. For this discussion, we will narrow our focus to the thalamus and the hippocampus, two subsystems of the limbic brain that have some of the closest association to the neocortex. Since the thalamus appears before the hippocampus in the processing chain of incoming sensory information, we will cover it first.

Thalamus

Communications Hub

The thalamus is a relatively small football-shaped structure centrally located within the brain. The neocortex curves around it in such a way that it is conveniently situated virtually equidistant radially from all neocortical regions. With the exception of our sense of smell, the thalamus is the only source of information about the body and its environment for the neocortex. It is well-suited for this task because of the resonant properties in the alpha brainwave band that exist between it and the neocortex, making it a key player in the determination of what information about the outside world, detected by our senses, can pass through and be distributed for further processing in neocortical networks.[103(pp177,184-186)] In fact, when there is a lack of environmental stimuli, it is this resonance that is ultimately responsible for the coherent alpha oscillations that arise in areas of the thalamus involved in the processing of sensory stimuli, and in the primary neocortical sensory areas that they are closely associated with.[103(p200)]

The thalamus does not appear to perform complex computations on any of the information passing through it because there isn't much coordination between adjacent neurons or subcompartments (thalamic nuclei). Despite the lack of its own inherent capacity to process information, the precisely wired resonant channels connecting it with the neocortex result in a combined thalamocortical system that is very good at carrying out the complex information processing required for higher mental faculties.[103(pp177,179)]

The neocortex is functionally segregated into distinct processing systems that each play a role in the processing of information ranging from first-order processing of sensory stimuli all the way to higher-level associations and executive functions. The first job of the thalamus is to route the incoming sensory signals to the appropriate areas of the neocortex. After accomplishing this first-order operation, it then helps route information from the primary sensory areas of the neocortex to those dealing with higher-order information, such as the intermediate and frontal levels of the neocortical hierarchy. Therefore, the thalamus can be considered as a communications hub that links large neocortical areas in a flexible manner. These links supplement the long-range links within the neocortex that connect distant assemblies, but since the thalamus is positioned nearly equidistant from all neocortical areas, it does so in a way that creates shortcuts and bypasses the progressively increasing axonal conduction delays in the direct long-range neocortical links.[103(pp185-186)]

Thalamocortical Circuits

Just like the neocortex, the thalamus has both excitatory and inhibitory neurons. Since there isn't much local computation occurring in the thalamus, the excitatory neurons send the vast majority of their axons to the neocortex; these projections are referred to as thalamocortical to indicate that input flows from the thalamus to the neocortex. Furthermore, these neurons release rhythmic action potentials similar to neocortical excitatory neurons and can be discharged by receiving excitatory input or by being released from inhibition.[5(p125),103(pp179,181)] The local inhibitory neurons in the thalamus have a different arrangement than those in the neocortex where they are close to most of their targets, nested within excitatory networks; in the thalamus, they are located in a thin shell called the reticular nucleus that surrounds the thalamic structure.

There are two sub populations of excitatory neurons that exist in the thalamus, and each type is a key participant in one of two types of circuits that exist between the thalamus and the neocortex (Figure 13a). The first category we'll call non-specific neurons. They form a network distributed throughout the entire thalamus and their thalamocortical projections connect the thalamus with all sensory and motor areas of the neocortex. Through the alpha brainwave oscillation, these neurons support the more global aspects of neocortical information processing by forming wide-spread connections in the superficial layers of the neocortex—mostly layer 1, but also layers 2 and 3. These layers are key zones for the "shell" of the minicolumns that populate the neocortex.[108,119] In addition to targeting superficial neocortical layers, non-specific thalamocortical projections also target layer 6, the deepest neocortical layer. The non-specific circuit targets large populations of neurons that often encompass several assemblies and by doing so, this circuit helps to temporally bind separate channels of information in the neocortex, making it possible for the unity of conscious experiences.[5(p125),124]

The second type of excitatory thalamic neuron is called specific neurons. These neurons cluster together in the thalamus, in what's called thalamic nuclei, and form a looping circuit between layer 5 or 6 neocortical pyramidal neurons. This type of thalamic neuron precisely targets the input layer of the neocortex, layer 4, where it can make contact with local inhibitory neurons, the apical dendrites of pyramidal neurons in deeper levels, or the spiny stellate neurons that then send signals to pyramidal neurons throughout the many layers of the neocortex. In addition to layer 4, specific neuron projections sometimes can target layer 5. In general, specific thalamocortical projections form strong, driving connections that branch profusely, resulting in a dense array of synaptic end points in layer 4.[5(pp125-126)] These driving projections are in contrast to modulatory ones like feedback or the projections of non-specific thalamic neurons to the neocortex. The central

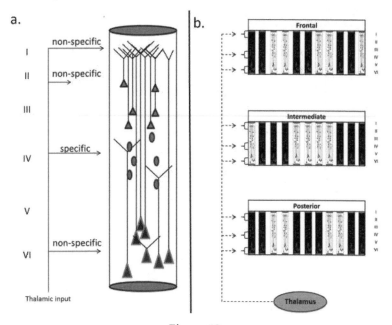

Figure 13

focus of the specific circuit is the layer 5 pyramidal neurons that form the "axis" of neocortical minicolumns and the principal job of this circuit is to hold neuronal activity over time at these locations, maintaining fast waves of activity and supporting more local aspects of neocortical information processing.[108,119] In short, specific loops maintain, or hold information content in one location over time.[124]

Waymarker #7: In B-U IF, thalamocortical neurons are analogous to dark matter and the Higgs boson; more precisely, the non-specific thalamocortical neurons correspond to dark matter, the specific thalamocortical neurons correspond to the Higgs boson.

It should be noted that although there are two distinct thalamocortical circuits, information from the two does, however, get mixed once it has arrived at the neocortex thanks to the inherent circuitry of the neocortex—i.e., there exists a variety of links between neurons in different neocortical layers.[105(p11)] In addition, unlike the pyramidal neurons in the deeper layers of the neocortex, layer 2/3 pyramidal neurons do not participate in strong thalamic loops.[108] The layer 2/3 pyramidal neurons send projections that tend to stay within the neocortex, targeting lateral areas or levels of the neocortical hierarchy below or above. Therefore, layer 2/3 pyramidal neurons, principal players in the non-specific and shell circuits, are involved in sending information throughout the neocortex while layer 5 pyramidal neurons, the principal players in the specific and axis circuits, specialize in maintaining activity over time, i.e., holding the information content.

There is close interaction between thalamocortical neurons, both specific and non-specific, and the inhibitory neurons that are distributed throughout the depth of the neocortex, but the interaction is most prominent in the layers that receive the bulk of the thalamocortical inputs, namely layers 1, 4, and 6. Inhibitory neurons in these layers are often directly targeted by thalamocortical projections, therefore, they are part of the immediate circuitry of thalamic input to the neocortex, giving them great influence over the receptive field properties of neocortical minicolumns.[107(pp65,92,97)]

The degree to which any one location within the neocortex interacts with the thalamus via the specific or non-specific circuits depends on the particular level of the neocortical hierarchy under consideration. On the lower levels of the hierarchy, where the primary sensory areas are, layer 4 is dense with stellate neurons who receive an abundance of specific circuit input. On the other hand, as we ascend the neocortical hierarchy to the intermediate levels, the ones that are involved in more advanced stages of sensory representation, we find that the density of layer 4 stellate neurons is reduced, indicating a reduction in the role of the specific circuit in layer 5 pyramidal neuron activity. This decrease in specific circuit input to intermediate levels of the neocortical hierarchy is accompanied by increased non-specific circuit input to those same areas, input that makes contact on the distal regions of layers 2/3 and 5 pyramidal neuron apical dendrites and tufts. This shift from specific to non-specific circuit input as we ascend the neocortical hierarchy is paralleled by a switch between bottom-up (feedforward) to top-down (feedback) activation. This could be due to the fact that stellate neurons play a critical role in the registration of incoming sensory information in the primary neocortical sensory areas, but once this has occurred, they are no longer needed in such high numbers, and so therefore, their densities decrease in the intermediate and higher levels of the neocortical hierarchy.[108]

Ultimately, layer 5 and 6 pyramidal neurons participate in both the non-specific and specific circuits between the thalamus and neocortex, but more information appears to flow from the circuits involving layer 6 pyramidal neurons to those involving layer 5 pyramidal neurons than the other way around. This is consistent with layer 5 pyramidal neurons occupying a central location in terms of both their position within neocortical minicolumns, and the flow of neocortical information. Layer 6 pyramidal neurons can provide input to layer 5 pyramidal neurons by either directly sending projections to their dendrites in layers 4-6, or, in the case of layer 6 thalamocortical pyramidal neurons, indirectly by making contact with specific thalamic neurons that, through the specific circuit, send projections to layer 4 stellate neurons who then send projections to the layer 5 pyramidal neurons. On the other hand, a relatively minor contribution of layer 5 pyramidal neurons to the layer 6 pyramidal neuron circuit occurs directly via layer 6 pyramidal neuron basal dendrites.[5(p126),108]

Subdivisions

I have identified three very general levels of the neocortical visual processing hierarchy—the posterior, intermediate, and frontal neocortical areas—and each are connected via specific and non-specific loops to different areas of the thalamus. The primary visual cortex, which deals with the early stages of visual information processing, is involved in a dynamic loop with an area of the thalamus that receives the incoming sensory information directly from the retina. The intermediate and frontal regions of the visual processing system are in a dynamic loop with areas of the thalamus that do not receive signals from the retina, rather, they deal with higher-order information received directly from the neocortex itself. In fact, although virtually all sensory information has to be routed by the thalamus to the neocortex, most of the input to the thalamus actually comes from the neocortex. The primary providers of this higher-level, associational input are layer 5 pyramidal neurons on the intermediate and frontal levels of the neocortical hierarchy who target thalamic areas not receiving primary sensory information. Furthermore, layer 5 pyramidal neurons provide driving input to the thalamus while layer 6 corticothalamic pyramidal neuron input tends to be more modulating.[103(pp177-178)] The higher-order thalamic nuclei are also critical sites for attentional processes because they receive feedback from frontal areas of the neocortex that can enhance or suppress thalamic activity, which would then subsequently affect the activity of the minicolumns in the intermediate areas of the neocortex that receive projections from these thalamic neurons.[105(p72),108]

Recall that I have defined a neocortical coalition to be a collection of cooperating assemblies that can span the neocortical hierarchy and are linked via long-range connections. The drawback to communication occurring along some long-range neocortical channels are the progressively longer axonal conduction delays (i.e., slowdown of signal transmission) that build up the farther the signals have to travel. Luckily, however, neocortical areas can also be linked via communication that comes in the form of thalamic relay neurons. Usually, the same neocortical areas that are connected via long-range direct neocortical links are also linked via cortico-thalamocortical shortcuts (from one neocortical area to a thalamic relay neuron, then to another neocortical area). Therefore, thalamic neurons assist in the combining of neocortical assemblies into a single large neuronal system—a coalition[103(p185),105(pp59,62,70)] (Figure 13b).

Hippocampus

Memory
Because its inputs are the explicit representations of all sensory modalities, the hippocampus is a prominent associative structure within the brain and is situated high up in the brain's processing hierarchy, closely associated with the intermediate

and frontal regions of the neocortex[5(p195),103(pp281,297),125,126,127] (Figure 14). Its most notable contribution to our mental abilities is the role that it plays in the formation of new declarative memories, or, memories that can be consciously recalled, such as episodes. Despite the importance of these memories and the fact that the hippocampus receives high-level sensory and episodic information to form them, the existence of consciousness surprisingly does not depend on a properly functioning hippocampus (and areas surrounding it in the medial temporal lobe such as the entorhinal and perirhinal cortices). This conclusion is supported by observations made of amnesiacs who have damaged hippocampal regions, and as a result, lack the ability to form new declarative memories. This does indeed have a profound negative impact on people with this affliction but they are still capable of being conscious.[5(pp194-195),125]

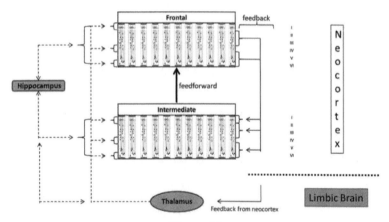

Figure 14

A key insight regarding the relationship between the hippocampus and the neocortex is that people with damage to their hippocampal regions are able to retain and use declarative memories that they have acquired before the damage occurred, indicating that the hippocampus is not the final storage location for the memories that we acquire. In a process that takes place over the course of a few weeks, the hippocampus transfers newly acquired episodes to the neocortex, mainly to the temporal and prefrontal lobes (intermediate and frontal levels of neocortical hierarchy, respectively), for consolidation with existing memories. It is in these locations that episodic-declarative memories can exist for long periods of time.[5(p195),125,128]

Structure and Function

Like the neocortex, the hippocampus contains both excitatory pyramidal neurons and inhibitory neurons. On the other hand, the layers that can be found in the two structures differ. However, enough correspondences can still be found to consider the hippocampus as a single giant neocortical module with largely random connections. This large random connection space provides the spatiotemporal context for information to be flexibly encoded using the synaptic strength between its neurons, making the hippocampus ideally-built for encoding external events and internal thoughts into temporary neuronal representations. In an attempt to illustrate the relationship between the hippocampus and the neocortex, neuroscientist György Buzsáki has described the hippocampus as being like the librarian to the library that is the neocortex—it is a subsystem of the brain that ensures information gets stored in a well-organized fashion.[103(Ch11)]

The large random connection space of the hippocampus allows for a practically unlimited number of trajectories that encode memories, and within this connection space, it is possible for signals to travel from one location to another along a variety of different routes, some involving direct transmissions while others involve numerous relay neurons. The exact trajectory of transmitted signals in the hippocampus depends on the strength of synaptic connections between neurons and the state of local inhibition; in other words, action potentials will travel along the path of least resistance—the path with the strongest synapses and weakest inhibition.[103(pp285,290-291)] The general plasticity rule that "neurons who fire together wire together" is ultimately responsible for the ability of the hippocampus to keep track of sequences of locations, events, and items.[103(p318)]

Recall that neocortical neurons predominantly communicate locally. In contrast, the storage of information in the hippocampus is spatially widespread, meaning that memory traces span large portions of the hippocampus. Within this random connection space, a single neuron is capable of communicating with the majority of the others, no matter if they are close by or on the other side of the hippocampal structure. Therefore, a collection of hippocampal neurons encoding a memory need not be located anywhere near each other in the hippocampus—this is an example of distributed coding.[103(p286)]

Waymarker #8: In B-U IF, the hippocampus corresponds to the source of the inflaton; its resemblance to a single, giant neocortical column, and the distributed coding that takes place within it, gives it properties that are analogous to what Inflation asserts, namely that somehow, all matter in the universe must have existed in a state where there was extensive interaction between all particles allowing for a uniform temperature distribution to be established before the sudden and dramatic onset of inflationary cosmic expansion.

In addition to temporary storage, the structural design of the hippocampus gives it a functional versatility ranging from error correction, pattern recognition, and signal amplification.[103(p285)] Furthermore, it can be considered as an autoassociator, or, a system that takes in input and recreates a pattern from the past that most closely matches the input. The neurons in the hippocampus undergo use-dependent changes that increase the probability that the same assemblies will be activated upon future presentations of the same stimulus, even in the event that the stimulus changes somewhat.[103(p247)] The result of these use-dependent changes is hippocampal attractor states that represent newly acquired episodic information and event sequences.[103(p301)]

Because it is an autoassociator, the hippocampus can aid in the brain's ability to make predictions by comparing information stored within it—memories—to incoming sensory information. This is a form of hypothesis testing and it affects the fate of incoming sensory information because the more accurate the hypothesis, the more widespread the sensory information will be forwarded from the hippocampus to regions of the neocortex.[103(pp323-326),110] The hippocampus is in a location where it can communicate with both the upper stages of the brain's sensory processing hierarchy and the prefrontal cortex, and it mainly sends its signals to these locations, allowing information about episodic events and relations to affect the content of consciousness and modify neocortical circuits, although, it remains true that a fully functioning hippocampus is not necessary for the manifestation of consciousness.[5(p196),103(p341),129] In turn, the hippocampus receives much of its input from the neocortex in the form of explicit representations produced by information processing that began in the primary sensory areas and then progressed through to the upper-stages of the processing hierarchy where neuronal activity tends to mirror our subjective percepts.[5(pp277-279),103(p281)]

The prefrontal cortex exists on the highest level of the neocortical hierarchy, in the frontal regions, and a location within it, the medial prefrontal cortex, is closely associated with long-term memory. This location shows greater activation than the hippocampus during the retrieval of remote memories while the hippocampus shows greater activation during the retrieval of recently acquired memories. The medial prefrontal cortex and the hippocampus are directly connected and communicate often during sleep when the newly acquired memories currently stored in the hippocampus are transferred to the neocortex for long-term storage.[130] In fact, during certain periods of sleep, the hippocampus communicates with various regions of the neocortex in such a way that it can make adjustments to synaptic strength. In addition to the prefrontal cortex, neocortical areas targeted by the hippocampus include cingulated, parietal, and visual cortices. Furthermore, correlated firing patterns observed during waking experiences are observed again during sleep between hippocampal neurons and neurons in these same neocortical regions.[130]

Hippocampal Rhythms

The characteristic brainwave frequencies of the hippocampus reside in the gamma and theta sub-bands of the brainwave spectrum. Gamma activity encodes the details of episodes, or, sequences of events and items, while theta waves are used to temporally organize the neurons and assemblies functioning at gamma frequencies. In other words, the theta waves modulate the gamma band activity of the neurons and assemblies involved in memory processes.[103(pp331,343),119,127,131]

Despite being comprised of a variety of different neurons and subcompartments, the hippocampus acts as a single oscillator that is very good at keeping time.[129] It uses the theta rhythm, which it is the primary generator of, to induce phase-locked discharge in places such as the areas that surround it (perirhinal and entorhinal cortices), prefrontal cortex, cingulated cortex, thalamus, and upper-stage sensory areas.[103(pp308,341),129] Because the hippocampus plays such a large role in the acquisition of new memories, it is not surprising that the theta rhythm is one of the more sustained rhythms observed in the awake and active brain, indicating constant interaction between the hippocampus and neocortex.[103(pp308,335)]

The theta rhythm is used as a multiplexing mechanism that can organize the timing of neocortical assembly oscillations.

Multiple hippocampal neurons, or neuronal assemblies, acting at the gamma frequency can be grouped onto a single theta cycle, allowing for the multiplexing of roughly 7 different items.[103(p321),119] These hippocampal neurons can then call up corresponding assemblies throughout the neocortex whose gamma oscillations are modulated by the theta rhythm.[103(pp340,351-352)] Neocortical assemblies can be made to maintain the same state over the course of a theta cycle, something that is useful for working memory, or assemblies can alternate activation throughout the theta cycle to represent an episode or sequence.

Map Making

The ability of the hippocampus to encode episodic information related to the outside world and our inner thoughts may have started out within the brains of early species of animals as a system that specialized in navigating physical spaces. There are neurons in the hippocampus that encode for locations in the environment, and collectively, these hippocampal "place" neurons form a cognitive map of the environment[103(pp296-297)] such that every path in physical space corresponds to a path in hippocampal neuronal space. As a consequence of the distributed information storing and processing of the hippocampus, it is synaptic strength between hippocampal neurons that encodes physical distances in the environment since the neurons in the hippocampus encoding for nearby locations in the environment need not be situated next to each other—the closer the two locations in physical space, the stronger the synaptic connection between the neurons encoding for them.[103(p299)] Corresponding to a route in physical space, the most effective route between any two neurons, even taking into account necessary detours, can be calculated within the hippocampal neuronal space.[103(pp301,329)]

The receptive fields of hippocampal neurons encoding for locations in the environment are learned, which can take anywhere from minutes to days.[103(pp298-299)] Furthermore, these receptive fields can change depending on environmental details, such as the size of the local area being encoded.[103(pp306-307)] The activity of hippocampal place neurons is organized temporally by the hippocampal theta rhythm, within which, multiple neurons become active to collectively encode an animal's position. The most active neuron corresponds to the one whose receptive field most closely matches the location of the animal. As the animal approaches the center of a neuron's receptive field, the neuron will begin to respond more vigorously; when the animal is leaving the center of a neuron's receptive field, the neuron begins to respond less vigorously.[103(p323)]

Hippocampal place neurons are also sensitive to the speed at which an animal is moving. They are known to possess what's called velocity-dependent gain which means that their discharge frequency increases the faster the animal moves. One of the results is that the number of action potentials emitted by the place neuron while the animal is in the receptive field of the neuron is held constant no matter how fast the animal traverses the receptive field. For example, if an animal traverses the receptive field of a neuron in 1 second, the neuron will be active in twice as many theta cycles than if the animal traveled at a faster rate and took only 0.5 seconds to travel through the receptive field. But to compensate for less time spent in the receptive field on the faster run, the place neuron will produce twice as many action potentials as it would if the animal took 1 second to traverse the same region of space.[103(pp323-324)]

As it turns out, there are aspects of map building that are analogous to encoding episodic memory and event sequences.[103(pp315-316)] The current view is that the hippocampus and its surrounding areas specialize in forming different types of maps, those related to our environment and those that keep track of events in our lives.[132] Collectively, they form a general system that encodes the information of both spatial and non-spatial relations—ordered locations and distances, and episodes and sequences of events.[103(pp301,328)]

Neocortical Hierarchy and Subsystems

The neocortex is an incredibly complex structure that is primarily responsible for the most advanced mental phenomena, like consciousness and self-awareness, possessed by some of Earth's life forms, such as mammals and primates. It is the most recent subsystem of the brains of mammals to have evolved but it does, however, work closely with systems within the limbic brain to accomplish its feats. Two structures that are closely associated with the neocortex include the thalamus and the hippocampus.

The thalamus is a centrally-located structure around which the neocortex curves, placing it nearly equidistant from all neocortical regions. In this location, the thalamus is able to act as a communications hub and provide shortcuts between neocortical modules that can circumvent the long-range neocortical connections and the axonal conduction delays that come along with them. Therefore, the thalamus is critical for the formation of neocortical coalitions whose assemblies can be widely distributed throughout the neocortex. Nearly all incoming sensory information is initially routed through the thalamus; however, there isn't much thalamic computation — it is the combined thalamocortical system that is capable of the complex processes responsible for advanced mental phenomenon.

The hippocampus is situated way atop the neocortical sensory hierarchy, in close proximity to the intermediate and frontal regions. It receives explicit sensory representations that it uses to construct sequences of places, events, and items — episodic memories — that it temporarily stores before eventually transferring to the neocortex for long-term storage. It is a prominent associative structure within the brain, but is not necessary for the manifestation of consciousness. The hippocampus has been likened to a giant neocortical module with a largely random connection space, enabling it to keep track of both spatial and non-spatial relations.

This chapter introduced you to neurons in the brain that are the analogs of the newly confirmed Higgs boson, and neurons that are the analogs to particles that are still considered theoretical, like dark matter particles, or particles that mediate cosmic inflation.

Having introduced these two structures that are closely associated with the neocortex, we will now consider in a bit more detail the hierarchical organization of the neocortex and the existence of specialized neocortical areas that carry out various aspects of the vast array of neocortical information processing. By doing this, we will be able to draw insightful conclusions about how consciousness arises within the brain.

Chapter 12

Intermediate-level Theory of Consciousness

Where Outer and Inner Worlds Interface

In a sense, the neocortex is a relatively homogeneous structure because all throughout it, there are six layers and modular functional units—minicolumns and columns; widespread features like these suggest that local computations in any neocortical area are fundamentally the same. Throughout the many different areas, however, neuron size and density does vary, as well as the connectivity between areas via long-range connections. The specific details of neuronal circuitry, such as afferent vs. efferent connectivity, ultimately determine the type of information that gets processed in a given neocortical area, such as whether it is visual or auditory.[103(p58)]

Within any given neocortical area, all of the neurons work together to form feature detectors.[5(p22)] For example, the visual system consists of a hierarchy of specialized areas that collectively perform a myriad of computations, some of which include: orientation identification, motion detection and direction estimation, edge detection, object detection and recognition, depth perception, facial recognition, determination of mood and mindset of partner in conversation, etc. Most of these computations are performed in parallel across a wide range of spatiotemporal scales and oftentimes involve filtering of signals and some form of multivariate analysis.[5(p211),102,105(p3),106(pp9,11,73)] It is estimated that the neocortical visual processing hierarchy of advanced primates consists of up to 32 areas with hundreds of feedforward and feedback interconnections,[107(p176)] but it should also be noted, however, that the borders between these areas are not always well-defined.[5(p119)] Furthermore, each distinct area can usually be broken down into several subsystems which is important to keep in mind because it means that the three very broad levels of the neocortical visual processing hierarchy that I have identified—the posterior, intermediate, and frontal levels—all consist of numerous individual areas who each contain their own subunits and compartments.[5(p121)]

Visual attention and awareness arise from the information processing, or, computations being performed by the neocortex. The neuronal substrate that directly represents the information underlying consciousness most likely resides somewhere between the explicit sensory representations of the incoming stimuli and the areas of the neocortex responsible for executive functions such as planning and decision-making. The explicit representations form in the intermediate levels of the neocortical hierarchy, which, for vision, occurs in and around the inferior temporal lobe. On the other hand, the executive functions are carried out in the frontal regions of the neocortex, such as the prefrontal cortex and the cingulated cortex.

Much of what is discussed in this chapter is based on Christof Koch's theory about how the brain produces consciousness which he has referred to as the Intermediate-level Theory of Consciousness. It suggests that feedback from the frontal areas to the intermediate ones is primarily responsible for producing our conscious experiences.[5(pp295-298)]

Hierarchy of Neocortical Areas

The response characteristics of neocortical neurons depend on which level of the neocortical hierarchy they reside. For instance, receptive fields get larger and their triggers more specific in the areas situated high in the neocortical hierarchy, indicative of the explicit representations that are present there. In addition, as one ascends the hierarchy, say from the primary visual cortex to the inferior temporal cortex, there is a shift from local operations that generate "maps"—point-by-point distributions of various quantities—to the integration of local information from different parts of the image, or from different operations carried out on the same location in the image. Thanks to this process, visible surfaces and their characteristics can be reconstructed, resulting in explicit representations.[5(pp137,151),106(pp14-15),108]

It should be pointed out that the visual processing hierarchy is defined based on anatomical considerations and not the exact timing at which the different areas receive signals after visual input. For example, not all areas on the same level of the hierarchy become active at the same time, and sometimes, like coprocessors, areas on different levels of the hierarchy can become active simultaneously. In short, activation does not always start out uniformly on the lowest levels of the hierarchy and then ascend it step-by-step through subsequent levels.[5(p123),105(pp126,398)]

The architecture of the neocortex makes it possible for the brain to combine the current sensory input with past experience and other forms of additional information. Initially, the presentation of stimuli produces feedforward signals that support a somewhat hierarchical analysis of the visual input. Then, well-defined waves of feedback signals from the front of the neocortex to the back can be observed about 200 to 400 ms after the presentation of the stimulus; this feedback is indicative of top-down aspects of visual processing such as the inferences and contextual information that the front of the brain uses to fill in for incomplete or ambiguous sensory information.[103(pp235-238),122] Bottom-up, feedforward signals are predominantly driving, meaning, they cause their targets to vigorously produce action potentials. On the other hand, most top-down, feedback signals are modulatory, although the strength of feedback seems to depend on the neocortical layer that it is destined for.[105(pp186,188,195)] Furthermore, because feedforward and feedback signals propagate along high-caliber axons and have similar conduction speeds, it takes the same amount of time for signals to either ascend or descend the neocortical hierarchy which makes it possible for neurons located on different levels of the hierarchy to simultaneously exchange information in both directions.[105(p196)]

Two paths

Within the visual system, there are two streams of information coursing through the neocortical hierarchy that begin to branch off from each other after the primary visual cortex, which is at the very back of the brain and on the lowest level of the hierarchy (Occipital lobe). The ventral stream, the one that takes the low route forward, is concerned with forms and object recognition—conscious perception—while the dorsal stream, the one that takes the route over the top of the brain, is concerned with several types of spatial information such as location, motion, and depth. The ventral stream is often referred to as the vision-for-perception stream, or the "what" stream, and the dorsal stream is referred to as the vision-for-action stream, or the "where" stream. The inferior temporal and parietal cortices are the two principal areas that exist on the intermediate levels of the visual processing hierarchy in the ventral and dorsal streams, respectively[133] (Figure 15).

Eventually, the two streams reunite in the prefrontal cortex, although, there are numerous links between them that appear along their ascent of the neocortical hierarchy and there are even areas that appear to be a part of both paths.[5(pp128-129)] Whether through the activity of the prefrontal cortex or the links joining the two paths, the brain is able to bind the features and objects from the ventral stream to one of the many spatial maps constructed by the dorsal stream, and there is evidence that if this does not happen, a complete conscious perception of a scene may not occur. For example, this has been observed in people with damage to their parietal region who can only perceive one object out of many, or one feature of an object at a time.[102,134]

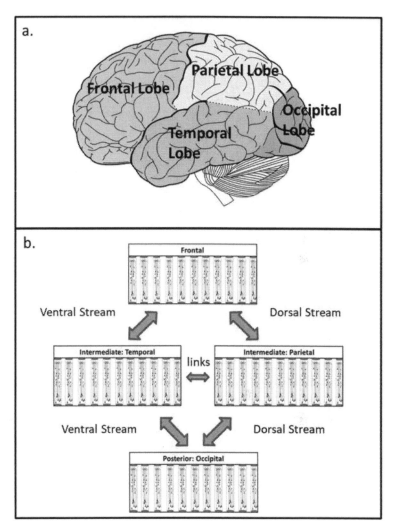

Figure 15

The ventral and dorsal streams can be traced back to parallel pathways that leave the retina and get routed through the thalamus. One pathway carries color and contrast information and is sent to the ventral pathway while the other carries spatial information, such as motion and depth, and is sent to the dorsal pathway.[107(p175)] Well-designed experiments allow for these pathways to be identified separately, but it must be pointed out that the two streams readily share information and that many of the functions that must be carried out by the visual processing system, such as depth perception, can be accomplished by using information from either stream.[107(p176)]

The information being processed in the ventral stream is vital to our conscious perceptions while the information in the dorsal stream seems to support many functions that are unconscious to us, such as the readying of the body to perform a movement, or the execution of reflex responses.[5(pp228-229)] The temporal activation patterns of neocortical areas in the ventral and dorsal paths differ since in the ventral path, areas seem to become active in a more sequential, feedforward manner, while areas in the dorsal path become active relatively early and simultaneously, which explains the fast reflex-like responses that result from dorsal stream activity.[105(pp208-209)] At the top of the dorsal stream, the parietal cortex generates many different types of spatial maps to support the various requirements of a conscious person able to initiate motor actions and interact with his or her environment. The requirements for conscious perception, namely, the ability to recognize an object no matter its size, orientation, or location, is fundamentally different from what is required to carry out motor actions based on vision, which includes knowledge of exact spatial relationships between the objects and the organism.[5(pp212,216)] Therefore, the spatial maps that support motor actions differ from the one that supports conscious perception.

Primary Visual System

Images resulting from visual stimuli are first registered by the retina and then sent to the thalamus to be routed initially through the lowest level of the neocortical visual processing hierarchy—the primary visual cortex. The projection of the visual scene onto the primary visual cortex is smooth and continuous on large scales but jittery and discontinuous on the scale of individual neurons and minicolumns[5(p78)] and in a sense, these functional units respond to particular locations within the scene much like pixels in a digital image.

The activity in the primary visual cortex is not directly responsible for our conscious perceptions. However, information encoded there does get sent forward to areas higher in the hierarchy where it can ultimately help to shape our visual awareness.[5(p106)] The explicit representations that exist in the primary visual cortex are of things like locations and orientations; the explicit representations of faces and objects do not appear until the intermediate levels of the visual processing hierarchy.[5(p86)] Activity in early visual systems largely reflects the images impinging upon the retina and we never become conscious of most of this information — factors such as attention and saliency determine the particular subset of this visual stimuli that ultimately makes it into our conscious perceptions.[5(pp113,275)] One of the main reasons early visual areas do not contribute directly to conscious experience is that they are not in close enough contact with the planning areas in the front of the brain.[5(p105)] However, by way of feedback from the inferior temporal cortex or the hippocampal region, memory information does get sent to the primary visual areas and can affect the ongoing processes there.[107(p463)]

Ventral Stream

Early visual areas, the ones on the posterior level of the visual processing hierarchy, map particular features across the entire visual scene, such as orientation, motion, depth, etc. These neurons, and the minicolumns that they form, encode for particular locations in the image, locations that are either in the periphery or in the central focus — the fovea. On higher levels of the visual processing hierarchy, the neocortical modules begin to react more to what is in the central focus of the retina. But even with this bias, the receptive fields of neurons in these areas encompass almost the entire visual scene—these neurons are still at least a little sensitive to what is in the periphery. The overall consequence of this foveal bias is that areas on the intermediate levels of the neocortical visual processing hierarchy no longer collectively map the entire visual scene like earlier areas. Furthermore, they receive extensive feedback from frontal areas which can communicate memory traces and attention signals that synchronize the activity of targeted assemblies and help coalitions gain dominance over competitors. Areas on the intermediate level of the visual processing hierarchy, like the inferior temporal lobe, are believed to be the locations where neuronal activity supports visual consciousness. In other words, if stimuli does not result in activity in and around the inferior temporal lobe, conscious visual perception does not occur.[5(pp137,148-149,286),105(pp185-186,346,360)] The columnar organization of the inferior temporal lobe—minicolumns and/or columns—explicitly represent the features of objects that we perceive, such as corners, geometric shapes, faces, spatial orientations, etc.[5(p149),105(p350)] While there may be neurons within the inferior temporal lobe that specialize in representing particular features, or particular objects, such as faces, in general, the representations that form in this region are probably more widely distributed and complex such that overlapping areas encode for multiple objects and object features.[5(pp150-151)]

As information ascends the primate neocortical visual processing hierarchy, the inferior temporal lobe is believed to be one of the last predominantly visual areas and it is near the apex of the ventral stream. It sends many of its projections to areas involved in memory, action, and planning, such as the hippocampal region and prefrontal cortex, and it also receives feedback from these areas. For example, the feedback that it receives from the prefrontal cortex is vital for consciousness since the prefrontal cortex is involved in executive functions such as planning and decision-making, and this feedback can help resolve the battle for inferior temporal real-estate between conflicting coalitions. In addition, the feedback from both the hippocampal area and the prefrontal cortex can load memories into the inferior temporal cortex.[5(pp148,284)] For instance, areas of the prefrontal cortex involved in short-term memory can explicitly represent, via elevated firing rate, several objects and their spatial relationships even when the stimulus is no longer present. These prefrontal neurons send feedback to inferior temporal areas, causing information related to previously seen stimuli to reemerge into consciousness. On the other hand, neuronal activity in the inferior temporal region is more transient when the stimulus presentation is also transient, suggesting that feedback from frontal areas is needed to maintain inferior temporal activity and produce a stable percept.[5(pp200,203,285),135] Activity produced by feedback from frontal areas to inferior temporal cortex, as well as spontaneously occurring inferior temporal activity, can also induce mental imagery which tends to be less vivid than the imagery produced by outside stimuli because early visual areas, those on the lowest levels of the visual processing hierarchy, are not involved; the feedforward input from neurons in early visual areas help to constrain the activity of the neurons on the higher levels of the visual processing hierarchy.[5(pp300-301)]

For complex scenes and objects, neurons in the inferior temporal cortex can utilize the discrete neocortical processing windows to temporally multiplex the different objects and their features to form a complex, unified representation of the scene. On the other hand, they can also multiplex current sensory representations of objects with memory traces of objects to create even richer experiences.[107(pp458-459)] This temporal coding scheme is an efficient way to segregate and organize the information of different objects, object features, and memory traces since all of these have overlapping neuronal representations in the inferior temporal cortex. A consequence of this is that it allows for current visual input and memories/associations related to the current input—both short- and long-term—to all be scrutinized together by the frontal regions and result in a unified conscious perception and interpretation of the current situation.[5(p260)]

Dorsal Stream

The dorsal stream contains regions that explicitly encode the speed and direction of moving objects, as well as depth information in the scene. A person with damage to areas within this stream will have trouble perceiving motion and spatial relationships, but may still be able to recognize the object that is moving since object recognition is handled by the ventral stream.[5(pp145,151,307)] The parietal cortex, which is at the apex of the dorsal stream in the back half of the brain, constructs several types of spatial maps, most of which relate spatial information to what is required for action; therefore most of these maps are used to carry out behavioral movements.[5(p145)] The parietal cortex also encodes the spatial relationships between objects, enabling us to holistically perceive scenes. Without this spatial perception map, we would still be able to perceive individual objects, or object features, but would lack the simultaneous perception of surrounding objects, or multiple features possessed by a single object.[5(pp146,184-185),134] Unlike the spatial perception map, we are not directly conscious of the spatial maps that are used to carry out our unconscious actions. In fact, the reason these actions are so efficient is that they take place unconsciously and do not have to wait for analysis to be performed by the frontal neocortical regions—conscious analysis of movements taps into a different spatial map, the one used to impose order on our conscious perceptions.[5(pp210-211)]

Many neurons in the parietal cortex can be considered both motor and sensory neurons since they send signals to both motor and visual areas in the frontal regions. Parietal motor neurons also send signals to subcortical areas like the spinal cord and brainstem.[5(p146)] In fact, many of the motor functions supported by these neurons can become so routine that they are carried out without significantly involving the frontal regions of the brain. In the beginning, however, during the learning phase, these motor actions require the attentional and planning resources of the front of the brain, but eventually, the dorsal stream, along with the basal ganglia and cerebellum, take over when the function has been sufficiently learned and becomes automatic through practice and rehearsal. This frees up the processing power of the front of the brain to focus on more novel experiences and plan for future concerns.[5(pp194,235-236)] Furthermore, stimuli can trigger feedforward activity in the dorsal stream that results in rapid unconscious behavioral responses. The actual awareness of motion in the visual scene and of reflexive bodily responses probably does not arise until feedback from frontal areas reaches the dorsal stream;[5(pp144-145,214,217),119] although the unconscious reflexive behavior precedes conscious perception, awareness of the action is soon to follow and the front of the brain can create the illusion that the behavior was consciously initiated.

By studying individuals whose parietal cortex is damaged, neuroscientists are able to gain insights about how we perceive motion and time. For example, the rare neurological disturbance called cinematographic vision reveals that in the brain, time is fundamentally discrete, no less than space which is encoded by neurons in the hippocampal region. In cinematographic vision, instead of seeing continuous motion, a person sees a flickering series of still frames in the frequency range of 6-12 Hz. This reveals that for the normal functioning brain, we should think of motion as "painted" onto each frame, or, processing period. In other words, motion is not experienced because of a change in position of an object between two consecutive processing periods, rather, it is due to the activity of neurons, such as their firing rate, during the individual processing periods defining

work with each other and with areas on the intermediate level of the hierarchy to maintain sensory and short-term memory information, retrieve long-term memories, and manipulate this data to consciously carry out executive functions such as planning and decision-making.[5(pp130,244),136] In particular, feedback from the prefrontal cortex, and other areas in the frontal lobe, to areas on the intermediate level of the neocortical hierarchy appears to play a critical role in conscious perception and recall. These feedback signals are capable of synchronizing the activity of neuronal assemblies, making it more likely that they will be part of a dominant coalition.[107(p294)] In addition, this feedback can play a role in attentional processes because it can be used to ready the lower levels of the visual processing hierarchy for expected input—this essentially represents the use of expectations to achieve more efficient processing of input.[103(p272)]

Regions of the frontal lobes are instrumental to the conscious recall of short- and long-term memories, both of which manifest in fundamentally different ways in the brain. Working memory arises due to sustained neuronal activity. For example, if someone is required to remember the identity of an object after it has been removed from view, activity of neurons in the prefrontal cortex will continue to encode for the identity of the object and its properties. On the other hand, long-term memory is associated with structural changes in the brain, or, changes to neuronal hardware, such as strengthened synaptic connections between neurons.[5(p188),103(p159)]

The output of the frontal lobe is primarily responsible for generating our conscious behavior and experiences. If the visual centers in the back half of the brain, the intermediate and posterior levels of the neocortical visual hierarchy, look at the outside world with the goal of forming explicit representations of features and objects within it, then the front half of the brain looks at the back and uses these explicit representations to think, form intentions, make decisions, etc. Key areas within the front of the brain, such as the anterior cingulated, prefrontal, and premotor cortices, send signals to the motor areas to generate conscious actions. On the other hand, these areas also send feedback to the top of both the ventral (inferior temporal cortex) and dorsal (parietal cortex) streams which results in sustained conscious visions.[5(pp298-299,304-305,324)] The frontal lobe is a complex and powerful computational entity, but we are not directly aware of all of the processes that go on there, most of which are a form of higher-level cognition wherein sensory information and abstract concepts and patterns are manipulated. We are only directly conscious of the sensory reflections of the higher-level cognition taking place in the front of the brain, such as the re-representation of some of this cognition in inner speech, imagery, and tones. This re-representation is made possible by feedback from the frontal areas to upper-stage sensory areas on the intermediate level of the neocortical hierarchy. Furthermore, this feedback represents just one of the possible interpretations that the front of the brain could have settled on given the multitude of internal mental and external sensory signals that it processes at any given time.[5(p296),7(p50)]

Waymarker #9: In B-U IF, the frontal and intermediate levels of the neocortical visual processing hierarchy correspond to the boundary and interior branes in the Holographic Universe framework, respectively. Recall that there are two types of feedback from the frontal level of the hierarchy to the intermediate level: feedback that precisely targets neurons in layers 6, 5, 3, and 2 of the intermediate level, and widespread feedback that largely targets layer 1; the first type of feedback corresponds to electromagnetic radiation (light) while the second type of feedback corresponds to gravity.

As we ascend the neocortical hierarchy, the characteristics of the pyramidal neurons begin to change. Pyramidal neurons high up in the ventral stream and in frontal neocortical areas tend to be large and have axons and dendrites that branch profusely, more so than pyramidal neurons at the back of the brain. They receive and integrate more excitatory input than their counterparts lower in the neocortical hierarchy, making them more powerful computationally.[5(p75),105(pp369-370,372,375)] Furthermore, there is evidence that the hierarchical organization observed in the back half of the neocortex, starting on the posterior level and extending through the intermediate levels, may in fact continue through frontal regions as well.[5(p123),136]

Gamma oscillations are used by certain parts of the prefrontal cortex to encode items in working memory[103(p245)] and theta waves can be used to chunk the gamma-encoded information together, a process known as multiplexing. This enables between 7 and 9 assemblies to be grouped together in a variety of ways such as the repeated activation of a sequence of assemblies in consecutive theta cycles, something that is useful for holding information in working memory. Alternatively, the activation of a set of assemblies can be alternated in a particular order throughout the course of a theta cycle, something that is useful for representing episodic memory.[103(pp352-353)] Furthermore, during visual working memory tasks, synchronized theta oscillations are observed between prefrontal areas and areas on the intermediate level of the visual processing hierarchy involved in visual perception.[138]

These brain rhythms help to achieve synchrony among groups of neurons which is an important aspect in the establishment of the dominant coalition encoding the content of consciousness. This is only necessary early on, however, because once dominance is established, the coalition can survive without synchrony for a finite amount of time since its competitors have already been suppressed. Feedback from the frontal lobe is closely associated with this synchrony, and hence, the establishment of a stable percept-supporting coalition.[5(p46)] However, even when a coalition is no longer the dominant one, its reduced activity can still have an impact on ongoing processes associated with more dominant coalitions.[5(pp253-254)]

It should also be pointed out that when the brain is not preoccupied with external stimuli, it enters a default mode where key areas in the frontal and intermediate regions work together to form a unified network, known as the default mode network. These key areas include portions of the prefrontal, parietal, and temporal cortices. When the mind is adrift, activity in this system is closely associated with both memory and imagination, processes that are used to review the past, as well as project into the future in a way that explores possibilities.[104,139,140]

Neuronal Correlates of Consciousness

With help from limbic areas such as the thalamus, hippocampus, and amygdala, the frontal regions and the highest levels of the sensory hierarchy in both the ventral and dorsal streams most likely re-enforce each other's activity, and the activity in earlier areas as well, forming a self-amplifying feedback loop, or, standing wave, between assemblies all throughout the neocortical hierarchy—a quasi-stable coalition.[5(pp97,253,261,324)] Visual consciousness is most closely associated with feedback from areas in the frontal lobes to areas in and around the inferior temporal cortex, the last stage of visual processing in the ventral—vision-for-perception—pathway. This feedback helps to maintain the activity of the assemblies there above threshold, and improves the odds that the coalitions which they are a part of will win out the competition against their competitors. In return, by maintaining vigorous activity, the coalition members in and around the inferior temporal cortex are able to forward their information to these same frontal areas which are involved in working memory and planning. As the focus of attention changes from object to object, or from feature to feature, the winning coalitions change as well.[5(pp173-174,203,305-306)]

The above discussion brings us to the activity principle which states that underlying every conscious percept is an explicit neuronal representation of what is seen that exists in the minicolumns and columns of the higher stages of the visual processing hierarchy. For any given conscious percept, there are many assemblies that represent the different aspects of the scene, and together, they form a coalition—a network of assemblies. If these assemblies project their information to areas in the frontal neocortical regions, such as the prefrontal and cingulated cortices, and then receive feedback from these same regions, they can become explicit representations and form the neuronal correlates of consciousness.[5(pp306-307)] In other words, conscious perception occurs when a burst of coordinated activity occurs between frontal regions and key areas on the intermediate level of the neocortical hierarchy, such as the inferior temporal and parietal cortices. Therefore, the activity on these upper stages of the ventral and dorsal streams represents the information supporting our consciousness; when activity in these areas fades, so do our perceptions.[5(p64),141]

The threshold for becoming aware of a slight stimulus at the edge of perceptibility varies from trial to trial depending on fluctuations in attentional and background brain states. The longer the coalition encoding for the stimulus stays above threshold and dominant, the more confident we are in what we perceive. Awareness so fleeting that we are not sure what we saw could be due to a briefly existing coalition encoding for the stimulus, or this coalition could have quickly lost the competition for dominance against other coalitions. Thanks to fluctuations within the spontaneous activity of the neocortex, the crossing of threshold for the coalition encoding for a barely perceivable stimulus is a probabilistic, or, random process.[5(pp254-255)]

In addition to stimuli at the edge of perceptibility, there are also stimuli that are ambiguous, or sensory stimuli that can be interpreted by the brain in more ways than one, such as visual illusions like the Necker Cube. Each interpretation has its own coalition competing against the others for dominance, and at any given moment, whichever coalition is the winner is the one that encodes the interpretation that we perceive. Over time, coalitions rise to, and fall from, dominance and our perceptions change accordingly. But at any given moment, there will only be one dominant coalition, and hence, one unique percept, a phenomenon known as the unity of consciousness. In each of the two cases discussed here, stimuli at the edge of perceptibility and stimuli that is ambiguous, it is not the stimulus that changes from trial to trial or over time, it is the state of the brain.[5(pp269-270)]

The coalitions that are sufficient for internally-generated mental imagery, which are generally less vivid percepts than those arising from normal seeing, are likely to be less widespread than the coalitions produced by external input, and may not reach all the way down to the lower tiers of the neocortical visual processing hierarchy.[5(pp305-306)] Furthermore, background consciousness, which typically consists of fleeting percepts, most likely is a result of transient coalitions. Focal attention, or, feedback from frontal areas, is required to sustain these coalitions and bring them to the forefront of conscious perception.[5(p261)]

Keep in mind that the neurons participating in the dominant coalition supporting conscious perception are also connected with many more neurons who are not actively participating in the winning coalition. These neurons encode information that is related in some way to what the winning coalition is encoding for, such as past associations, expected consequences, cognitive background, future plans, etc. Because they are not directly participating in the dominant coalition, these neurons most likely do not fire vigorously, but they may have an elevated resting voltage level that brings them closer to threshold, and as coalition dominance changes over time, neurons and assemblies that were once on the periphery of the dominant coalition could become part of the new dominant coalition. A dominant coalition's connections with neurons on the periphery are important because they help to provide meaning to the current content of consciousness.[5(pp242,309-310)]

Attention

Attention is one of the hallmarks of the waking state and the two types that operate in parallel in the brain are close associates of consciousness. On one hand, when top-down volitional attention is applied, the front of the brain selects coalitions encoding for features, objects, or internal thoughts to enhance, improving their chances of achieving or maintaining dominance. On the other hand, bottom-up, saliency-based attention is made possible by feedforward signals capable of rapidly producing dominant coalitions. If top-down attention does not reinforce these coalitions, they can rapidly fade and give way to other coalitions. Top-down attention is useful for forming coalitions whose assemblies are not yet well connected; in this case, the front of the brain uses this type of attention to reinforce the activity of assemblies that are attempting to encode for relatively novel stimuli, enabling them to forward their signals up the processing hierarchy and reach areas involved with planning, short-term memory, and awareness. Saliency is explicitly coded for in frontal regions of the brain and is used to decide where the searchlight of top-down attention will get focused. On the other hand, extremely interesting — salient — features within the environment are able to "pop" into our consciousness without the immediate need for top-down attention.[5(Ch9,pp308-309),142]

The inferior temporal cortex plays a large role in the attentional processes necessary for identifying an object in a scene cluttered with distractor objects.[5(pp177-178)] Even when the target object is extremely salient and is presented alongside much less salient ones, the coalitions encoding for the less salient objects will eventually dominate within the inferior temporal cortex, even if for only a brief period of time and without being boosted by top-down attention.[5(pp271-272)] In general, however, if more than one object is placed within the receptive field of a neuron in the inferior temporal cortex, it will respond to all of the objects, but without vigor due to cancellation effects and inhibition. Under this type of condition, it is hard for the neuron to forward its information to higher levels of the neocortical hierarchy where it can contribute to coalition formation and dominance. On the other hand, top-down volitional attention, which is mediated by feedback from frontal areas, helps to determine which object the neuron ends up encoding for, and in the process, suppresses the activity dedicated to encoding for the objects not being attended to—the distractor objects.[5(pp176-177),108]

Top-down volitional attention is capable of affecting neuronal activity on all levels of the neocortical hierarchy, as well as in the thalamus, but it does not do so equally. It is primarily generated in the frontal areas, on the highest level of the neocortical hierarchy, and from there, the attentional signals are fed back strongly to the next level below, the intermediate level. On the other hand, attentional signals take longer to reach the posterior level of the neocortical hierarchy and are much weaker there.[5(pp178,180)] In general, top-down attention synchronizes the activity of coalitions and improves their chances of becoming dominant. Alternatively, if external sensory input is strong enough, it can reset the phase of ongoing oscillations to induce synchronization which may be enough to produce a transient dominant coalition.[103(p255)]

The Isolated Mind

In this chapter, we considered in more detail the hierarchical organization of the neocortex. To accomplish this, we focused largely on the visual processing system which is the largest of the sensory systems and spans every level of the neocortical hierarchy. The primary visual cortex exists on the posterior level of the neocortical hierarchy and is the location where visual input is initially registered inside the brain. From there, two streams of information travel throughout the neocortex. The ventral stream travels the low route forward through the temporal cortex and is concerned with explicit representations of objects and their features; hence, this path is closely associated with conscious visual perception. The dorsal stream takes a route over the top of the brain through the parietal cortex, is concerned with motion and depth information, and is responsible for reflexive movements that are largely unconscious. Both the temporal and parietal cortices exist on the intermediate level of the neocortical hierarchy and both forward their information to the frontal lobe where executive functions take place. Ultimately, several areas within the frontal lobe work together during memory and attentional processes, decision-making, emotional and linguistic expression, etc. But perhaps the most important function of the frontal lobe is to provide the feedback to the intermediate areas that produce our conscious experiences by forming and fortifying dominant coalitions. This feedback can also be interpreted as the sensory re-representation of a subset of the higher-level, "supra-conscious" sensory processes that take place in the frontal lobe, i.e., just one of the many possible interpretations produced by the front of the brain.

You should find that the discussion in this chapter concerning levels of neocortical hierarchy connected via feedforward and feedback projections evokes the M-theoretical concepts in chapter 6 of branes connected through an 11th dimension to form systems of branes.

For the most part, we have been largely concerned with the brain while it is in the waking state. But when we go to sleep, and our brain becomes almost entirely isolated from our environment, it enters a variety of other states, some simple, some complex,

sensory or cognitive information and begins the internally-driven evolution through states where it systematically interacts, or, communicates with structures closely associated with it. As NREM sleep progresses through its stages, we enter deeper and deeper sleep which means that the waking threshold increases the farther we advance into NREM sleep. In stage 1, the brain transitions from awake to sleep and the neocortex displays low-voltage brainwaves in the slow alpha and theta frequencies. During stage 2 sleep, the thalamus begins to communicate with the neocortex by sending it signal bursts known as k-complexes and sleep spindles. During stage 3 sleep, communication via spindles becomes more prominent and is now orchestrated by neurons all throughout the neocortex that begin to synchronously cycle at delta frequencies through periods of activity and inactivity called up- and down-states, respectively. In addition to spindle-mediated communication between the thalamus and the neocortex, there are also signals being communicated from the hippocampus to the neocortex in the form of high-frequency packets known as sharp-waves and ripples, both of which are also orchestrated by the up-down oscillations of neocortical neurons. In stage 4, spindle activity almost stops entirely, leaving the delta activity as the dominant feature. The fifth stage is known as REM sleep and it is when your typical semiconscious dreaming occurs. The neuronal activity in the neocortex during this phase of sleep displays a similar mixture of gamma, beta, alpha, and theta frequencies as the waking state, which is not surprising since we know that dream experiences can often be as real as waking ones. Furthermore, the waking threshold is lowest during this stage, making it the shallowest phase of sleep. The stages where spindling is prominent, stages 2 and 3, are also known as slow-wave sleep because of the delta frequency oscillations. This type of sleep accounts for approximately half of all sleep; dream sleep on the other hand accounts for up to a quarter.[103(p187)]

At the conclusion of stage 5, a new cycle begins. Generally, four or five of these NREM-REM cycles occur in the average complete night of sleep, each lasting between 70 and 90 minutes. However, as sleep progresses, the characteristics of the cycles begin to change. Early in the night, sleep is dominated by deep sleep—stages 3 and 4. During this time, REM sleep—stage 5—is very short. By the middle of the night, stage 4 may no longer manifest while stage 3 continues longer into the night through about 2/3 of the total period of sleep. From the middle of sleep through to the end, the REM periods begin to lengthen and dominate sleep. With the shift from the dominance of stages 3 and 4 to the dominance of stage 5, sleep gradually becomes shallower as the night progresses[103(p187)] (Figure 16).

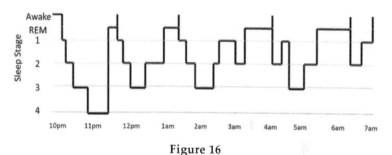

Figure 16

Generally, the types of memories that get processed during NREM and REM sleep are different. The sources of memories during NREM sleep are episodic memories while the dreams that occur in REM sleep have high emotional content and involve procedural memories, or, memories for the performance of particular types of actions. In addition, the brain simultaneously records memories of events that were consciously perceived and those that were not, but there is a preference, however, for memories related to consciously perceived events to be processed during sleep.[128,137]

NREM Sleep

The principal accomplishment of NREM sleep is the processing of memory information. Repeated reactivation of neuronal circuits and replay of firing sequences during restful waking periods and NREM sleep is necessary because it allows for the molecular processes involved in strengthening and weakening synaptic connections to take place.[103(p208)] Neurons that became active during recent activity—recent meaning earlier in the day or even during the last week or so—participate in the NREM memory processes. All of this activity is triggered by the up-down cycling of neocortical neurons whose synaptic strengths have been modified by waking experiences.[103(pp194-195),125] Since both the thalamus and hippocampus were highly active during the initial experience, these areas also display reactivation and replay during NREM sleep. Recall that the hippocampus temporarily stores recently acquired memories and eventually ships them to the neocortex for long-term storage and consolidation with existing memories, a process that, for any one memory, can take up to three weeks. The replay and reactivation of neurons throughout the hippocampus and the intermediate and frontal levels of the neocortical hierarchy is integral to this process.[125]

We tend to be in a non-conscious state during NREM sleep because the memory processes occurring at this time, such as the consolidation of new memories with those in long-term storage, are carried out by the same neurons that are involved in the processing of conscious sensory and cognitive information; these neurons are enslaved to the massive self-organized NREM memory processes, and therefore, cannot simultaneously engage in the tasks necessary for the type of conscious perception experienced during the waking state.[103(p188),128] There are advantages to performing memory consolidation during off-line states, like the ability to compress large amounts of information within the critical window of synaptic plasticity, and the ability to reduce interference between new memories and existing long-term memories.[128] Although NREM is the period of time when the brain works on our memories the most, some memory consolidation can still occur while we are awake and conscious but only during non-attentive states.

After being awakened during NREM sleep, it's usually hard for people to remember what they were just experiencing, but in the somewhat rare cases when they do remember their experiences, they usually report experiencing some sort of thought or sensation that is different in character from the hallucinatory experiences of a typical REM sleep dream. For the average individual, NREM is a non-conscious state, however, there does appear to be a small percentage of people who are naturally able to retain some degree of self-awareness while in this collection of sleep states.[143,144]

In addition to people naturally gifted when it comes to NREM self-awareness, there are those who, through a tremendous amount of practice, are able to develop an ability to perceive mental states during NREM. A great example would be people who have expertise in transcendental meditation; when examined in sleep labs, many of these individuals show above average ability to achieve some degree of lucidity and/ or altered brainwave spectral characteristics during NREM.[143,145]

The vast majority of scientific observations of this type of NREM self-awareness have only verified it in the shallower stages of NREM sleep, namely stages 1 and 2. Furthermore, it appears harder to manifest NREM self-awareness the further one progresses into NREM sleep, that is, into stages 3 and 4.[143] It's interesting to note that, as stated earlier, stages 3 and 4 are the deepest stages of sleep, the ones with the highest waking threshold. So taken together, the insight from observations of both NREM and REM lucidity are suggestive of an inverse relationship between the ease of induction/amount of sleep lucidity and the depth of sleep. In other words, it's easier to induce high levels of lucidity in shallow sleep stages (NREM stages 1 and 2, and REM sleep) than it is in deeper stages (NREM stages 3 and 4).

Spindles

Spindles are highly coherent and widespread activity throughout thalamocortical circuitry that last for approximately 4 seconds and can be observed in both the neocortex and the thalamus during slow wave sleep.[146,147] Although the thalamus is the principal driver of spindle activity, it is the neocortex that determines where and when the thalamus will send spindle signals. For example, neocortical down-up cycles at the delta frequency provide a trigger to the thalamus to emit spindles, and neocortical long-range connections ensure that spindle initiation occurs synchronously.[146] At least part of the control that the neocortex has over thalamic spindle initiation involves contact with thalamic inhibitory neurons because high levels of inhibition are associated with delta frequency thalamocortical oscillation, while lesser amounts of inhibition result in oscillations at spindle frequencies.[148] The selective aspects of neocortical control over spindle activation depend on the waking experiences that leave their mark on the synaptic strengths between neocortical neurons. During the slow oscillations of NREM sleep, these changes determine which neocortical and thalamic neurons participate in the next spindle event. Through this process, neurons that were active during waking experiences can become reactivated at night during the processing of memories[103(p209)] and the amount of spindles is positively correlated to the amount of learning that occurred during waking experiences.[149]

Spindles begin to appear sporadically in stage 2 sleep, then peak in stage 3, and finally begin to taper off in stage 4. They mainly dominate the intermediate and frontal levels of the neocortical hierarchy,[103(pp200-201)] and in some cases, there are two different varieties of spindles that target these two neocortical areas — spindles higher in frequency (14 Hz) and more rhythmic target the intermediate areas while the lower frequency (12 Hz) and less rhythmic variety target the frontal areas.[147]

Sharp-waves and Ripples

During waking experiences, prominent hippocampal activity occurs at rhythmic theta frequencies. During slow wave sleep, however, irregularly-occurring, brief, and high-frequency activity begins to appear. These are large-scale hippocampal oscillatory patterns that last only tens to hundreds of milliseconds and come in at least two forms, sharp-waves and ripples. Ultimately, this activity is involved in the communication of information to the neocortex for memory consolidation, i.e., the downloading of recently acquired memories held in the hippocampus to the neocortex to become long-term memories. These brief transmissions contain highly compressed memory traces, or, sequences that originally spanned several seconds during waking experiences, but are now condensed into the punctuated sharp-waves and ripples.[125] During these short bursts of hippocampal activity, neurons are reactivated and sequences of their firings are replayed. Since the hippocampus sends signals to the neocortex, replay occurs there as well, such as in the visual centers in temporal and parietal cortices and in regions of the frontal lobe involved in memory and motivation. It is also possible for the hippocampus to communicate with the neocortex in this way outside of slow-wave sleep, but only during non-attentive waking periods.[125,128,130,146,150] Therefore, this type of hippocampal activity is an example of internally-driven self-organized activity within the brain because it occurs when there is virtually no interaction with the environment, allowing for repeated replay of neuronal sequences modified during waking experiences.[103(pp344-345)]

Similar to the relationship between the neocortex and

thalamus during spindle activation, it is the down-up slow oscillations of the neocortex that call up the hippocampal neurons to generate the sharp-waves and ripples. From that point, it is the hippocampal burst initiators that induce the replay observed in several regions of the neocortex.[128] When learning occurs during the waking state, the hippocampus gets stimulated, leading to changes in the synaptic strength between the responding hippocampal neurons. This ultimately is responsible for the sharp-waves and ripples generated during slow-wave sleep because neuronal activity is going to follow the path of strongest synaptic connectivity.[103(pp348-349),151]

"Sharp-wave/ripple" events can involve up to 100,000 hippocampal neurons which make them one of the most synchronous network patterns in the brain. This large-scale hippocampal event is capable of affecting large neocortical areas distributed throughout the intermediate and frontal levels of the neocortical hierarchy, a quality that gives the sharp-wave/ripple event the ability to transfer large amounts of information to the neocortex in the process of memory consolidation.[103(p345)] When hippocampal information is sent to the neocortex during these times, it is compressed into short bursts, allowing for episodes to be replayed multiple times and for information to be combined, creating new associations. During waking experiences, events that occurred outside the window of synaptic plasticity do not get connected synaptically to create associations, but since more information can be brought within this critical window during sharp-wave/ripple events, these associations are now possible. Furthermore, when the hippocampal information generated during sharp-wave/ripples is sent to the neocortex, associations between it and the information already existing in long-term storage can now be formed.[103(pp346-347)]

Spindle and Sharp-wave/ripple Co-occurrence

During slow-wave sleep, the activity of just about all excitatory and inhibitory neurons in the neocortex is synchronized by the slow oscillations between highly active up-states and completely inactive down-states. When the state of the neocortex transitions from a down-state to an up-state, it can send signals to burst initiators in the thalamus that trigger thalamic spindles. The result is spindle activity in thalamocortical circuitry that lasts for approximately 4 seconds with a frequency of approximately 12-14 Hz. In addition, the neocortical transitions from down-states to up-states can send signals to burst initiators in the hippocampus to trigger sharp-waves and ripples. This results in short-lived and synchronous high frequency transmissions of memory traces back to the neocortex. The co-occurrence of spindles and sharp-waves/ripples, which peaks during stage 3 sleep, creates the optimum conditions within the neocortex for solidifying temporarily stored memories that have recently been acquired.[128,130] Thalamic spindles put their target neocortical neurons into the best state to undergo changes in the strength of their synaptic connections which are then adjusted according to the hippocampal sharp-wave/ripple input. In this way, sharp-wave/ripples replay memories within the thalamocortical circuitry of the intermediate and frontal levels of the neocortical hierarchy that have been reactivated by spindle activity.[130,131,146] The brief hippocampal sharp-wave/ripple (anywhere from 30-200 ms) occurs at the onset of thalamocortical spindles which last for approximately 4 seconds, during which time it is expected that the activity of spindling neocortical neurons is biased towards adjusting their synaptic strengths based on the hippocampal memory traces provided as input at the very beginning of the spindle event.[103(pp350-351),146] Different memories can be processed in successive cycles of spindle and sharp-wave/ripple events since the particular thalamocortical and hippocampocortical circuits that are active in each cycle are variable.[131]

Waymarker #10: In B-U IF, the co-occurrence of spindles and sharp-waves/ripples within thalamocortical and hippocampal circuits corresponds to the current state of the universe; spindles are the analogs of the Higgs ocean and the dark matter cocoon that encapsulates normal matter; sharp-wave/ripple input to the neocortex is the analog of the inflaton; the shift of the neocortex from a down-state to an up-state is analogous to the Big Bang.

REM Sleep

Recall that the stages of deep sleep—those with the highest waking thresholds—are in NREM sleep. REM sleep, on the other hand, is a much more shallow form of sleep just below the waking threshold. In fact, many people often briefly wake up or enter a very light stage of sleep upon completion of a REM cycle before restarting another full cycle through the stages. The fact that REM sleep occurs near the waking threshold and is also the sleep stage where semiconscious dreaming occurs is reflected in the spectral content of the brainwave activity during this time, namely, the brainwave spectrum during REM sleep resembles brainwave activity occurring during waking experiences. Almost the entire neocortex becomes active in this way during dreams, with the exception of critical areas involved in self-reflection in the prefrontal cortex and early visual areas on the lowest level of the neocortical visual processing hierarchy such as the primary visual cortex.[5(pp99,109),136]

The extensive communication of information from the hippocampus to the neocortex that was observed during NREM sleep no longer exists during REM sleep, indicating that the role of REM sleep in the processing of memories is different from the redistribution and consolidation of recently acquired memories. Acting in absence of hippocampal input during REM, the activity of the neocortex during dreaming is more indicative of adjustments in synaptic strengths designed to explore potential associations;[152] the lack of hippocampal input means that neocortical activity can be more random and less restricted. In addition, the dorsolateral prefrontal cortex, critical for our self-awareness and critical reflection, is typically not active during REM, further freeing neocortical activity to explore what would normally be viewed as irrational associations.[130,136,137]

The information being processed during NREM and REM periods differ, but are still complementary nonetheless. While the focus of NREM sleep is the transfer of episodic memories to neocortical areas, the content of dreams do not involve much episodic information other than the fact that they are constructed from discrete elements of episodes that get combined to form associations in a relatively unrestricted fashion. On the other hand, allowing REM sleep to immediately follow NREM sleep improves the retention of procedural memories—those that are involved in the unconscious performance of actions.[136,137] In addition, dreams play an important role in the processing of emotional memories. In fact, the inactivity of the dorsolateral prefrontal cortex during normal dreaming can be attributed to increased activity of emotional processing centers such as the amygdala.[137]

In contrast to the non-conscious state that we are typically in during NREM sleep, we are in a semiconscious state when dreaming in REM sleep. This is a secondary type of consciousness as opposed to the primary consciousness which manifests during active engagement with the environment and/or thoughts. The activity of many frontal areas that are involved in reasoning and planning is responsible for the secondary consciousness of dream sleep; however, the lack of primary consciousness is due to the inactivity of the dorsolateral prefrontal cortex, an area of the frontal lobe responsible for critical reflective awareness. It is, in fact, possible for the activity level of the dorsolateral prefrontal cortex to vary during a REM period so that as its activity picks up, we experience more reflective awareness. This type of increased awareness is associated with a phenomenon called a lucid dream in which the dreamer realizes that he or she is dreaming while still within the dream.[136]

Mind Models ALL

We have seen that when we are not actively engaged with our environment or thoughts, our brains begin to wander through self-organized trajectories of attractor states. By far, the most prominent of such default modes occurs when we go to sleep at night, during which time the brain is practically isolated from the environment and body, and there is no directed thought. The complex trajectory that manifests is vital to a healthy functioning brain and the states that arise can be classified into five distinct stages of sleep. The first four can generally be considered as non-conscious states and are collectively known as NREM sleep because semiconscious hallucinatory dreaming does not occur; on the other hand, the fifth stage—REM sleep— is when the typical semiconscious state of dreaming does occur. The processing of episodic memories is the focus of NREM sleep while procedural memories, emotional processes, and free associations are explored during REM sleep. Each complete cycle through these states lasts for roughly 90 minutes, and during the average complete night of sleep, the brain will execute 4 or 5 such cycles. As the night progresses, there is a shift from NREM sleep dominance to REM sleep dominance.

Stage 1 is the transition from the waking state to sleep. In stage 2, the thalamus communicates to the neocortex via bursts of signals called k-complexes and spindles. In stages 3 and 4, neurons begin delta frequency oscillations between down-states of complete inactivity and up-states of high activity globally throughout the neocortex. These oscillations orchestrate neocortical communication with the thalamus and hippocampus. The switch from "down" to "up" triggers 4-second long spindles within thalamocortical circuits that were active during recent waking experiences. At the same time, the up-state can trigger sharp-waves and ripples in hippocampal neurons. Both spindles and sharp-waves/ripples play an important role in memory processes, such as consolidation, because spindles reactivate thalamocortical circuits while sharp-waves/ripples reactivate hippocampal and neocortical neurons and cause them to replay firing sequences from the waking state. The brief sharp-wave/ripple input (~100 ms) to thalamocortical circuitry at the beginning of a spindle event makes it possible for the strength of neocortical connections to be adjusted based on the episodic information stored in the hippocampus.

This chapter prepares you to glimpse in great detail the information processing event—concurrent spindle and sharp-wave/ripple events—that occurs in our brains when we sleep that I assert is analogous to the current state of our universe.

Here, in the second part of this book, the goal was to outline a conceptual model of neocortical and limbic aspects of the human brain. Next, in the third part of the book, I will explicitly state parallels that I see between this system and the system presented in Part I—the universe. These correspondences have much potential to illuminate the way forward for the theorists attempting to solidify the leading theory emerging in physics today, string theory. Furthermore, these correspondences can provide evidence-based answers to many of the classic philosophical questions concerning the nature of reality, the purpose of life, and the existence of gods. In short, the correspondence between the human brain and the universe is the principle upon which a true theory-of-everything will ultimately stand.

Part III

Correspondence

Chapter 14

Correspondence

The Model Within

I have now reached the point where I can use the models presented in Parts I and II to explicitly state some of the correspondences that exist between the structural organization and dynamics of the universe and the human brain. Before I begin, however, let me state a bit more precisely than I have thus far what aspects of the human brain I will be focusing on, because as we have seen, it is a tremendously complex system capable of manifesting a wide array of states. I will primarily concern myself with the subsystem formed by the neocortex, thalamus, and hippocampus—what I'll refer to as the thalamocortical-hippocampal system. Furthermore, I will compare the *current* state of the universe to this particular subsystem when the brain is in a very specific state, namely, during stage 3 of NREM sleep, when it is cycling through inactive and active periods, the latter being when spindling and sharp-waves/ripples concurrently manifest. During this time, the lowest level of the neocortical hierarchy, where the primary visual cortex resides, is largely silent since thalamic spindles mainly target the intermediate and frontal levels of the hierarchy, and the hippocampus-generated sharp-waves and ripples are most directly sent to these same areas.

If the universe and the human brain are isomorphic to each other, then they can be used to model each other. While pointing out parallels, I will alternate between the two directions of information transfer: (1) when knowledge of the human brain can inform theoretical physicists on the quest to mathematically model the universe at the most fundamental level, and (2) when knowledge about the universe can illuminate the way forward for neuroscientists trying to crack the neocortical code. Throughout this chapter, I will draw links between some of the more well-defined concepts within theoretical physics and neuroscience, showing a particular interest for what theoretical physicists can possibly learn about some of the more unsettled issues within their field if they decide to use the characteristics of the human brain as a guide. A good place to start is with Figure 17 where an M-theoretical view of the universe is shown alongside a diagram of the areas of the brain that we are concerned with; organizational similarities between the two systems can immediately be identified.

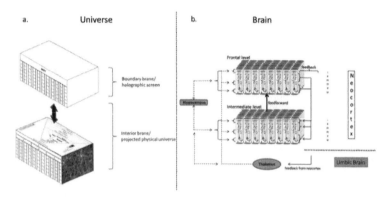

Figure 17

Layered and Modular

Recall from Part I that in one particular type of universe allowed by M-theory, spacetime is computed by 10-dimensional objects called branes within which string vibrations produce all of the particles in the Standard Model, and hence, all physical matter. Nine of the brane's dimensions are spatial and one is time; of the nine spatial dimensions, three are the extended dimensions that we are directly conscious of, the other six are extremely tiny and curled up at every point in space, too small even for our most powerful sensors to detect. The exact geometry of these tiny dimensions is not well known but believed to be a particular class of shapes known as Calabi-Yau shapes. In fact, one can think of the 10-brane that computes spacetime as a giant array of Calabi-Yau shapes; their six extra spatial dimensions can be considered as tiny and curled up, and the geometry of the Calabi-Yau modules can be considered to be determined by the details of the fluxes existing between the dimensions. There are practically an unlimited number of Calabi-Yau shapes that are possible and a major challenge facing theoretical physicists is identifying the one(s) that describes the universe we live in; the geometry has to be one that allows strings to vibrate the matter of our universe into existence and one that gives spacetime its other properties, such as the correct rate of cosmic expansion.

The neocortex has the same properties as the branes of M-theory. The activity of its neurons encodes the four-dimensional character of our environment—the same four extended dimensions that are in string/M-theory. It has both a layered and modular design with its six layers stacked one on top of the other and its fundamental module, the minicolumn, arrayed all throughout. Just like the six extra spatial dimensions in an M-theoretical universe extend the dimensionality of classical spacetime, the six neocortical layers extend the dimensionality, or degrees-of-freedom, of neocortical information processing beyond just the four extended dimensions that we already know it encodes. Furthermore, neocortical circuitry facilitates fluxes from one layer to another, much like the fluxes between the six curled up dimensions that determine a Calabi-Yau shape's geometry. Note that a similar argument regarding degrees-of-freedom in neocortical information processing was provided by György Buzsáki,[5(p58)] however, he only sited the five degrees-of-freedom that result from the distribution of unique types of excitatory neurons throughout layers 2 through 6. I, on the other hand, am suggesting that all six layers, including layer 1 where only inhibitory neurons and the distal apical tufts of pyramidal neurons are present, account for six degrees-of-freedom in that the ultimate Holographic Universe theory that will be found to pertain to our universe will be one where even the boundary brane is fully described by the fundamental entities of M-theory—strings, branes, extra dimensions, etc.; therefore, Figure 17a depicts both the interior and the boundary branes as being 10-dimensional.

Before turning our attention to the brain with regards to it containing a two-tiered, hierarchical substructure, let's briefly revisit the resonance between the Holographic Universe framework and quantum mechanics. Recall that in quantum mechanics, all physical systems are associated with a nonphysical probability wave that encodes all the information for the states that the system could manifest if sufficiently measured by its environment; this probability wave is considered a purely mathematical construct and is not believed to be a real physical entity, only the outcomes of measurements, or, the physical system that results from the probability wave's collapse is capable of being observed. In a Holographic Universe, the physical universe exists within the interior brane, a projection of the processes occurring on the boundary brane. Since the interior is where what we perceive as physical processes occur, then perhaps the boundary brane can be considered to be the location of "nonphysical process" such as the probability waves of quantum systems; because the processes that occur on the boundary and in the interior are so tightly interlinked, the boundary brane is a viable location for these nonphysical quantum processes to occur. The collapse of a system's probability wave, which causes it to manifest one out of its possible states, is similar to the process of illuminating a hologram to produce a holographic projection. Furthermore, the nonlocal nature of quantum systems is compatible with the distributed nature of holographic information storage on a boundary brane.

Now, let's consider the neocortex. It too has a hierarchal

organization; its posterior, intermediate, and frontal areas constitute three levels of a hierarchy that work together to perform unified neocortical information processing. Since we are concerned with the neocortex during the spindles and sharp-waves/ripples occurring in NREM sleep, we can omit the posterior level of this hierarchy because it is largely silent at this time. Therefore, we end up with a two-level hierarchy similar to the one in the Holographic Universe framework (Figure 17b), and just as the two branes in the Holographic Universe can interact through a dimension connecting them, the frontal and intermediate levels of the neocortical hierarchy are able to interact via feedforward and feedback projections. So in a sense, it can be said that neocortical information processing has the same dimensionality—11 dimensions—as the Holographic Universe; it encodes time and the three extended spatial dimensions that we perceive, consists of six layers, and has a hierarchical structure linked via feedforward and feedback projections—a total of 11 dimensions.

Now let's go beyond just structural similarities between a Holographic Universe and the neocortical hierarchy and consider the functional relationship between the levels within both hierarchical structures. In the Intermediate-level Theory of Consciousness, it is the feedback from frontal areas to intermediate ones that results in primary consciousness. In fact, the activity of neurons in frontal areas reflects aspects of our mental processes that we are not directly conscious of; what we are conscious of are the largely feedback-mediated sensory re-representations of a subset of these processes. In this type of system, some of the information present in the frontal areas never manifests in consciousness, and some things that we were once conscious of, but have since slipped out of our consciousness, are no longer represented in the intermediate levels but may still produce activity in the frontal areas, as in short-term memory. This information can re-enter consciousness at a later time similar to how quantum systems can once again manifest physically if sufficiently measured, or how a hologram only produces its projection so long as it is illuminated. Therefore in this analogy, feedback from frontal to intermediate neocortical areas results in primary consciousness like the boundary brane projects to the interior, resulting in the physical universe. I would like to point out here that although I stated earlier that our main focus is on the thalamocortical-hippocampal system while it is in the non-conscious state of NREM sleep, it is still useful to consider aspects of the waking-life relationship between the frontal and intermediate areas, namely how feedback from the former results in the neuronal activity in the latter that directly underlies the content of our consciousness.

Consistent with my above assertions is the recent trend in psychology to use a quantum mechanical framework to more accurately model aspects of human cognition, such as decision-making, memory, reasoning, and perception.[18,19,20] That is, rather than modeling decision-making using classical probability theory, psychologists are starting to employ concepts from quantum mechanics, such as indefinite superpositions of states and measurement-induced collapse of these states, to more accurately describe how the human brain functions. Applying the mathematics of quantum mechanics to human cognition is proving to be an approach that better captures the ability of our brains to function in uncertain and ambiguous situations.

Neocortical Circuitry, the Standard Model, and Beyond

The exact details of how neurons are distributed throughout the neocortical layers, their physical characteristics such as dendrite types, and their inter- and intralaminar axonal projections determine neocortical circuitry, and hence, the fluxes that are possible throughout its six layers and between the different levels of the neocortical hierarchy. Since I am making the claim that minicolumn circuitry is analogous to the particular Calabi-Yau shape that determines the geometry of the universe, and hence, the string vibrations that are possible at each point in spacetime, it should be possible to see a resemblance in minicolumn circuitry and the relationships between the particles in the Standard Model, since it is, after all, theorized that string vibrations produce these particles. Furthermore, because the graviton and other forms of matter and energy, such as the Higgs and dark sectors, are also theorized to be manifestations of string vibrations, we should be able to find their analogs within thalamocortical-hippocampal circuitry as well. We shall now take a closer look at correspondences that exist between the circuitry and dynamical processes of the thalamocortical-hippocampal system and a universe theorized by string/M-theory (Table 5).

Universe	Brain
up-quark	layer 5 pyramidal neuron (in upper half of layer 5)
down-quark	layer 5 pyramidal neuron (in upper half of layer 5) in a highly synchronous circuit with a layer 6 corticocortical pyramidal neuron
strange-quark	layer 5 pyramidal neuron (in bottom half of layer 5)
charm-quark	layer 5 pyramidal neuron (in bottom half of layer 5) in a highly synchronous circuit with a layer 6 corticocortical pyramidal neuron
bottom-quark	long layer 6 pyramidal neuron (claustrum projecting cells)
top-quark	long layer 6 pyramidal neuron (claustrum projecting cells) in a highly synchronous circuit with a layer 6 corticocortical pyramidal neuron
electron	layer 2/3 pyramidal neuron (in middle or upper third of layer 2/3)
muon	layer 2/3 pyramidal neuron (deep in layer 2/3)
tau	layer 2/3 pyramidal neuron (deep in layer 2/3, border with layer 4)
electron-neutrino	layer 4 stellate neuron (bottom third of layer 4)

Table 5a

Universe	Brain
muon-neutrino	layer 4 stellate neuron (middle third of layer 4)
tau-neutrino	layer 4 stellate neuron (upper third of layer 4)
Standard Model particle generations	neuron depth within layer (also topological "cavities", or "holes" in minicolumn circuitry)
antiparticle	antidromic stimulation (axoaxonic connection near axon terminals)
charge (color and electromagnetic)	phase influences that layer 2/3, layer 5, and some layer 6 pyramidal neurons have on each other with the help of local inhibitory neurons
gluon	axosomatic/axoaxonic connections between pyramidal neurons in layers 5 and 6 (mediated by inhibitory neurons)
$W^{\pm}$ and Z^0 bosons	layer 6 corticothalamic pyramidal neuron (two types based on the targets of axonal projections)
photon (virtual kind underlying static electric fields)	axodendritic connections between layer 5 pyramidal neurons, layer 6 pyramidal neurons, and layer 2/3 pyramidal neurons (all of which can be mediated by inhibitory neurons)
electron energy levels	axosomatic connections between the vertically-arranged layer 2/3 pyramidal neurons within the same minicolumn (mediated by inhibitory neurons)
photon (real kind comprising EM radiation)	feedback produced by incoherent activity in frontal level of neocortical hierarchy that targets the dendrites of layer 5, layer 6, and layer 2/3 pyramidal neurons in intermediate level of neocortical hierarchy

Table 5b

Universe	Brain
graviton	feedback produced by synchronous activity in frontal level of neocortical hierarchy that targets layer 1 of the intermediate level of neocortical hierarchy
Higgs boson	thalamocortical pyramidal neuron (specific)
dark matter	thalamocortical pyramidal neuron (non-specific)
dark energy	axonal conduction delay in lateral axonal projections
inflaton	hippocampal input to neocortex during sharp-waves and ripples
superpartners (Supersymmetry)	inhibitory neurons
Calabi-Yau shape	minicolumn circuitry
4 extended spatiotemporal dimensions	the thalamocortical-hippocampal system's natural ability to represent the 4 extended spatial dimensions
6 tiny curled-up extra spatial dimensions	6 neocortical layers within a minicolumn
11th dimension added by M-theory	feedback and feedforward projections between frontal and intermediate levels of neocortical hierarchy

Table 5c

Universe	Brain
string vibrations	current flow/oscillations in minicolumn circuitry
boundary brane	frontal level of neocortical hierarchy
interior brane	intermediate level of neocortical hierarchy
superposed states	coherent representation of simultaneous attractors (for the same "percept") that exist on frontal level of neocortical hierarchy and who do not produce feedback to intermediate level or to the thalamus
collapse of wave function	one attractor on frontal level gains dominance (perhaps as a result of perturbation in the form of input from surrounding neurons) and is sent via feedback to the intermediate level and the thalamus
nonlocality	a single attractor underlying any given percept communicated from frontal level to intermediate level of neocortical hierarchy; the unity of consciousness
black hole	highly synchronous, vigorously oscillating coalition of neurons on the frontal level of the neocortical hierarchy that produces feedback solely destined for layer 1 of intermediate level
Big Bang and colliding branes	neocortical shift from "down-state" to "up-state" during NREM sleep
Higgs ocean, dark matter cocoon, inflation, and the start of universal evolution	co-occurrence of spindles and sharp-waves/ripples
localized increases in complexity throughout the universe	initial coherence and synchrony at the start of a spindle and sharp-wave/ripple event giving way to increasing levels of complex dynamics over time in neocortical locations receiving the input

Table 5d

Layer 5 pyramidal neurons correspond to the lighter quarks of the first two generations in the Standard Model, that is, either an up-quark or a strange-quark, depending on the neuron's depth within layer 5. It is known that layer 5 pyramidal neurons tend to be arranged in two clusters of roughly three neurons each, and each cluster is situated at a different depth within layer 5 — one higher, one lower.[153] The higher cluster corresponds to up-quarks; the lower cluster corresponds to strange-quarks. The larger pyramidal neurons in layer 6, the claustrum projecting pyramidal neurons, correspond to the lighter third generation quark, that is, the bottom-quark. If either of the above neurons enters a highly synchronous circuit with a layer 6 corticocortical pyramidal neuron so that both neurons are "tightly bound", what results is a configuration that corresponds to the heavier quarks of a generation, that is, down-, charm-, or top-quark.

When any of these neurons are in the same minicolumn, the influences that they are able to exert directly onto each other's somas and axon initial segments, which is mediated by inhibitory neurons,[154,155,156] correspond to the emission of gluons by quarks. In other words, these transmissions that help to tightly bind the activity of layer 5 pyramidal neurons, large layer 6 pyramidal neurons, and layer 6 corticocortical pyramidal neurons into circuits within a minicolumn correspond to the strong nuclear force that tightly binds quarks to form other composite particles, such as protons or neutrons.

Because the strong nuclear force can sometimes be repulsive and sometimes attractive, depending on the color charge of the interacting quarks, layer 5 pyramidal neurons and large layer 6 pyramidal neurons will have to show the same variety of interaction. Note that inhibitory neurons mediate the influence that layer 5 and layer 6 pyramidal neurons have on each other's somas and axon initial segments, the latter being the location where action potentials are generated. Nearly all synapses on pyramidal neuron somas and axon initial segments are from inhibitory neurons,[103(p66)] however, pyramidal neurons can provide stimulation to these inhibitory neurons that causes them to then send signals to the soma and/or axon initial segment of other pyramidal neurons. The interesting thing is that one of these inhibitory neurons, the one that targets the axon initial segment, has special properties that may allow for interactions analogous to the strong nuclear force. Although this is an inhibitory neuron that we are talking about, it has been found to actually have the ability to also depolarize, or, excite the targeted pyramidal neuron axon initial segment depending on that neuron's membrane potential. When the targeted pyramidal neuron is at resting membrane potentials, the inhibitory neuron excites it; when at elevated membrane potentials, the inhibitory neuron inhibits it.[154,155,156] To see how this can produce interactions analogous to the strong nuclear force, first consider what would be an attractive interaction between layer 5/6 pyramidal neurons. Direct excitatory axoaxonic stimulation from one of these pyramidal neurons to the axon initial segment of another (albeit mediated by an inhibitory neuron) can produce attractive interactions because the stimulation supplied to the targeted pyramidal neuron's axon initial segment pushes it closer to, or beyond, the threshold for action potential initiation, which ultimately moves the phase of its oscillatory/spiking activity closer to the phase of the stimulating pyramidal neuron. The attraction in this sense is between the oscillatory/spiking phase of the layer 5 and large layer 6 pyramidal neurons. On the other hand, a repulsive interaction would require the stimulating pyramidal neuron to provide the soma or axon initial segment of the targeted pyramidal neuron with inhibitory input (made possible by a mediating inhibitory neuron), thus, pushing the oscillatory/ spiking phase of the targeted pyramidal neuron further away from the phase of the stimulating pyramidal neuron.

For neuroscientists searching for novel insight regarding communication between the neurons discussed above, particularly when they are within the same minicolumn, the communication between quarks who combine to form, say, protons and neutrons, via the strong nuclear force, should be considered since the neurons and the action potentials they trade correspond to quarks and gluons, respectively. Recall that during strong interactions, the color charges of quarks and gluons make it so that when quarks trade gluons, the quarks swap their color charges. Might the neurons discussed above and the action potentials that they trade possess a property analogous to quark and gluon color charge? Perhaps color in the visual sense, something that we already know can be encoded by neocortical pyramidal neurons and their action potentials, is this analogous property. This prospect is even more intriguing when one considers the fact that 1) the brain employs trichromatic color vision, and 2) three color charges are required to describe quark interactions via the strong nuclear force in the interior brane of the Holographic Universe. If such a correspondence exists between neocortical trichromatic color vision and the color charges underlying the strong nuclear force, it would make for a truly amazing set of circumstances, a huge hint at the deep connection between the brain and universe.

Recall that quarks can convert into each other by way of the weak nuclear force. During this process, a $W^{\pm}$ boson is emitted which quickly decays into the two lepton particles of a Standard Model generation. Within B-U IF, this process corresponds to either the formation or termination of circuits between layer 5 or layer 6 large pyramidal neurons and a layer 6 corticocortical pyramidal neuron. Every time one of these circuits is formed or terminated, a layer 6 corticothalamic pyramidal neuron receives depolarizing input which quickly causes it to send signals along an axoaxonic projection up to layers 4 or 2/3 where it targets stellate neurons and/or layer 2/3 pyramidal neurons. This makes layer 6 corticothalamic pyramidal neurons analogous to the $W^{\pm}$ bosons, layer 4 stellate neurons analogous to neutrino$_e$, neutrino$_\mu$, and neutrino$_\tau$, and layer 2/3 pyramidal neurons analogous to electrons, muons, and tau particles. Consider the case when a down-quark converts to an up-quark by emitting a W^- boson that quickly decays to an electron and an anti-neutrino$_e$ (Figure 18). This corresponds to the termination of a circuit between a layer 5 pyramidal neuron and a layer 6 corticocortical pyramidal neuron. During this process, either one of these neurons, or both, sends a signal to a layer 6 corticothalamic pyramidal neuron (Figure 18a-b) that then sends a signal up to layer 2/3 where it makes an axoaxonic synapse on a layer 4 stellate neuron that synapses with a layer 2/3 pyramidal neuron. This type of projection from the layer 6 corticothalamic pyramidal neuron is capable of simultaneously depolarizing the layer 2/3 pyramidal neuron and the axon of the layer 4 stellate neuron

(Figure 18c-d). Exciting the axon of the layer 4 stellate neuron can excite its soma antidromically (indicated by the dotted axon of the layer 4 stellate neuron, as opposed to the solid axons of the conventionally-excited neurons in the figure; the same format is used for Figures 19 and 20). Recall that antidromic stimulation implies that an action potential is made to travel in the reverse direction along a neuron's axon, and if this action potential were to encounter a conventional action potential, one generated by the initial segment of the neuron's axon, the two action potentials would annihilate each other upon impact.

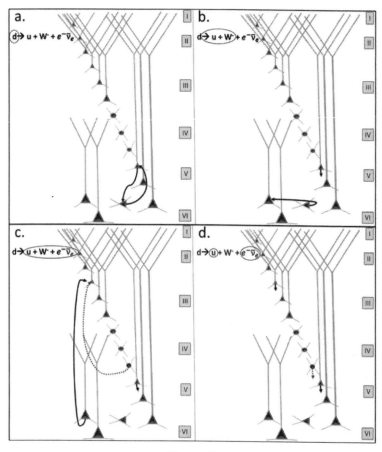

Figure 18

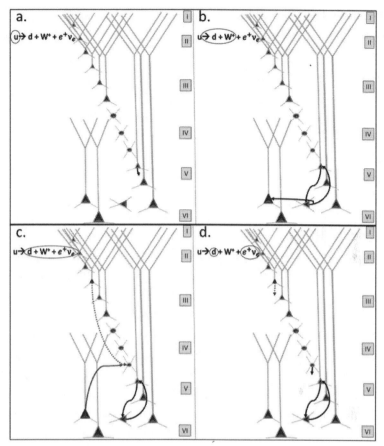

Figure 19

This process is analogous to the relationship between matter and antimatter. Therefore, in B-U IF, a layer 4 stellate neuron stimulated antidromically corresponds to an anti-neutrino$_e$.

Now consider the case when an up-quark converts to a down-quark by emitting a W^+ boson that quickly decays to an anti-electron and a neutrino$_e$ (Figure 19). This corresponds to the formation of a circuit between a layer 5 pyramidal neuron and a layer 6 corticocortical pyramidal neuron, and again, during this process, either one of these neurons, or both, sends a signal to a layer 6 corticothalamic pyramidal neuron (Figure

19a-b) that, this time, sends a signal up to layer 4 where it makes an axoaxonic synapse on a layer 2/3 pyramidal neuron that synapses with a layer 4 stellate neuron. This type of projection from the layer 6 corticothalamic pyramidal neuron is capable of depolarizing the layer 4 stellate neuron and the axon of the layer 2/3 pyramidal neuron (Figure 19c-d). Exciting the axon of the layer 2/3 pyramidal neuron excites its soma antidromically, which would make it the analog of an anti-electron.

$W^{\pm}$ bosons mediate weak nuclear interactions that do not preserve the charges of the particle undergoing transformation. In the above examples, the negatively-charged down-quark transmutes to the positively-charged up-quark, and vice versa. But recall that the weak nuclear force can also mediate processes in which charge is conserved. The mediating particle in these cases is the Z^0 boson. In B-U IF, this corresponds to a second type of layer 6 corticothalamic pyramidal neuron whose projections to layers 6, 5, 4, and 2/3 target pairs of identical kinds of neurons, or pairs of identical sets of neurons. Consider the case when a Z^0 boson decays into a neutrino$_e$ and an anti-neutrino$_e$ (Figure 20). This corresponds to the second type of layer 6 corticothalamic pyramidal neuron sending an axoaxonic interlaminar projection up to layer 4 where it targets the lateral projection of one layer 4 stellate neuron with an adjacent one (Figure 20a); the one that is depolarized in the usual fashion corresponds to a neutrino$_e$, the one that experiences antidromic depolarization corresponds to an anti-neutrino$_e$ (Figure 20b).

Using layer 6 corticothalamic pyramidal neuron circuitry to encode processes analogous to weak nuclear interactions suggests two patterns of interlaminar and intralaminar projections for these neurons: one neuron—the one that corresponds to $W^{\pm}$ bosons—primarily targets connections (axoaxonally) between neurons in different layers, like projections from layer 2/3 to layer 4, and vice versa; the other corticothalamic neuron— the one that corresponds to Z^0 bosons—targets connections

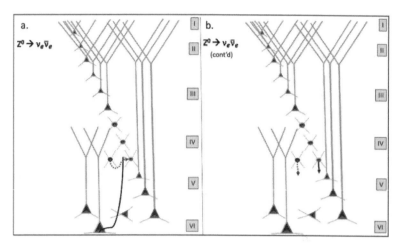

Figure 20

(axoaxonally) between neurons in the same layer, like lateral projections between adjacent neurons in layer 2/3, layer 4, etc. If layer 6 corticothalamic pyramidal neurons possessed inter- and intralaminar projections that formed axoaxonic synapses with their targets, they could function as "gate keeper" neurons who enable and facilitate communication within the neocortex. However, I am unaware of these neurons ever being observed to have these qualities. With that said, it should also be pointed out that neuroscientists are far from being able to characterize every single connection between all the neurons within even a tiny piece of brain the size of a "pinhead".[116] What this suggests is that there definitely appears to be room for discovery when it comes to the types of connections that exist within the neocortex.

We have just seen that in B-U IF, there is a relationship between: 1) the depth of pyramidal neurons in layer 5 and layer 6, and 2) the Standard Model generations. As one descends from the higher cluster of pyramidal neurons in layer 5, to the bottom cluster in layer 5, then down to the large pyramidal neurons in layer 6, the corresponding generation of quark goes from first, to second, to third. B-U IF says that similar patterns exist for the correspondence between other neocortical neurons and Standard Model particles. The stellate neurons that are positioned low in layer 4 tend to be smaller and send focused projections deeper into layer 3 where the pyramidal neurons are analogs of the electron. Middle layer 4 stellate neurons are larger and their projections begin to be more diffuse and terminate lower in layer 3, closer to the border of layers 3 and 4 where there are larger pyramidal neurons who are the analogs of the second generation lepton — the muon. Lastly, very large stellate neurons in the upper regions of layer 4 send widely diffuse projections to the border of layers 3 and 4 where large pyramidal neurons are the analogs of the third generation lepton — the tau. This makes the stellate neurons in the lower, middle, and upper regions of layer 4 the neutrino$_e$, neutrino$_\mu$, and neutrino$_\tau$, respectively. Pyramidal neurons in layer 2 and the upper regions of layer 3 correspond to electrons. Their particularly prominent vertical arrangement corresponds to electron energy levels where the bottom position in this arrangement is the ground state and each level up is an increase in excitation.

B-U IF says that the connections between neurons on the same level of the neocortical hierarchy mediate interactions that are analogous to the forces currently included in the Standard Model — the strong and weak nuclear forces, and the electromagnetic force. I have already outlined how this could be the case with the two short-range nuclear forces. Recall that I suggest that in the neocortical processes analogous to the strong interaction, inhibitory neurons serve as an intermediary and provide both excitatory and inhibitory impulses. Similarly, I suggest that inhibitory neurons will once again play an important role in neocortical processes analogous to the electromagnetic interaction because the electromagnetic interaction also comes in two varieties — repulsive and attractive. When it comes to activity analogous to the electromagnetic force, it is lateral axodendritic connections between pyramidal neurons in layers

2/3, 5, and 6, with some help from local inhibitory neurons, that can exert a "push" or "pull" on the oscillatory phases of the targeted neurons, similar to the repelling electric force felt between two particles with the same charge, or the attractive electric force felt between two particles with opposite charge. For example, in B-U IF, layer 2/3 pyramidal neurons correspond to negatively-charged leptons, like the electron. Therefore, the lateral connections between layer 2/3 pyramidal neurons should mediate a phase "push" to each other, which would require first making contact with an inhibitory neuron that targets the dendrites of layer 2/3 pyramidal neurons. The situation gets a bit more complicated for interlaminar connections between layer 2/3 pyramidal neurons and the deeper neocortical layers because the neurons there can be analogs of both positively- and negatively-charged particles. When a layer 2/3 pyramidal neuron is in communication with a circuit comprised of layer 5 and/or layer 6 pyramidal neurons that corresponds to one of the positively-charged configurations, the communication occurs through direct axodendritic excitatory input. On the other hand, when a layer 2/3 pyramidal neuron is in communication with a circuit comprised of layer 5 and/or layer 6 pyramidal neurons that corresponds to one of the negatively-charged configurations, the communication is mediated by inhibitory input to the dendrites of the neurons involved.

The interlaminar and intralaminar connections on both the frontal and intermediate levels of the neocortical hierarchy correspond to the forces in the Standard Model, but the connectivity on these two levels does differ from each other, such as in the complexity of the interactions they produce. Recall that in many ways, the frontal level of the neocortical hierarchy is more complex than the levels below. Likewise, theoretical findings within M-theory suggest that the boundary brane—analog of the frontal level of the neocortical hierarchy— has a more complex set of interactions than the interior brane—

the analog of the intermediate level.

The feedback that connects the frontal and intermediate levels of the neocortical hierarchy mediates interactions analogous to electromagnetic radiation and gravity. Recall that photons emitted by a charged particle, or system of particles with a net charge, whose state does not change are the virtual kind that mediate the static electric force. When the state of the charged system of particles changes, it will emit a real photon. In the framework being developed here, a charged object corresponds to a coalition of neurons spread throughout the intermediate and frontal levels of the neocortical hierarchy. When the state of the coalition changes, the portions of it on the frontal level can communicate these changes via signals sent along axodendritic lateral projections within the frontal level. When this "disturbance" encounters other active neurons who are themselves a part of other distinct coalitions of neurons, it can induce these frontal level assembly members to send feedback destined for layers 2, 3, 5, or 6 of the intermediate level, feedback that corresponds to real photons, or, light emitted out into the environment of a source object. Since layers 5 and 6 correspond to high-energy process, such as quarks and nucleons, the feedback that targets these layers must correspond to high-energy photons, such as gamma or x-rays. On the other hand, since layer 2/3 pyramidal neurons correspond to electrons, the feedback targeting these layers must correspond to lower-energy radiation, such as visible, IR, microwaves and radio waves.

The much more widespread feedback from the frontal level of the neocortical hierarchy to layer 1 of the intermediate level corresponds to gravity. Furthermore, the way the frontal pyramidal neurons generate this type of feedback appears to be complementary to the way it produces the feedback discussed above, namely, feedback destined for layers 2, 3, 5, and 6. I assert that this latter type of feedback is generated when a coalition axonal conduction delays experienced by these signals when transmitted over long neocortical distances reduce the rate of their transmission; neurons on the receiving end of these signals will detect a signal of lower frequency than what was transmitted, similar to the redshifted, or, the reduced frequency of the light traversing the expanding voids of the universe. In the analogy, voids correspond to neocortical regions that are not actively participating in the current spindle and sharp-wave/ripple event, regions that are not targeted by thalamocortical spindles; the long-range lateral projections along which the electromagnetic analogs are transmitted pass through these regions and connect neocortical regions who are actively participating in the current offline memory processing session.

The specific and non-specific thalamocortical projections correspond to the Higgs boson and dark matter, respectively. Recall that during waking experiences, virtually all sensory information is routed through the thalamus before reaching the neocortex, and during NREM, only the neocortical neurons targeted by spindling thalamic neurons are active during the concurrent spindle and sharp-wave/ripple events that I say correspond to our universe. This gives thalamic neurons the properties that make them vital for neocortical activation, analogous to the roles of both the Higgs boson, the fundamental particle theorized to be responsible for conferring mass to the particles in the Standard Model, and the dark matter that, via gravitational interactions, provides the scaffolding for the large-scale structure of the universe. By making direct contacts in layer 4 where stellate neurons play a role in the dissemination of the input to pyramidal neurons throughout all neocortical layers, specific thalamocortical projections provide the fundamental input that allows areas of the neocortex to participate in the ongoing information processing. Furthermore, during NREM sleep, only those neocortical areas receiving spindle input become active to participate in neocortical activity; therefore, the spindle input at this time corresponds to the Higgs ocean that first appears at the very beginning of the universe, just after the Big Bang.

Non-specific thalamocortical input targets neocortical layers associated with the shell of minicolumn circuitry, namely, the superficial layers and deep down in layer 6. They target layer 1 with widespread projections that can make contact with inhibitory neurons and the apical tufts of pyramidal neurons in layers below; because of these projections, this thalamocortical circuit is able to coordinate and isolate the activity of large populations of neurons. In B-U IF, these layer 1 projections correspond to a form of gravity. Therefore, the projections that non-specific thalamocortical neurons send to neocortical layers 1 and 6—layers that in B-U IF play a large role in processes analogous to gravity and the weak nuclear force, respectively— gives non-specific thalamocortical neurons properties that are analogous to those believed to be held by a particular type of theorized dark matter, namely, the kind that interacts with normal matter only through the force of gravity and the weak nuclear force (WIMPS—weakly-interacting massive particles).

Supersymmetry can be accounted for by the network of inhibitory neurons that are located in the thalamocortical-hippocampal system and are complementary to the networks of excitatory neurons in these same areas. In the neocortex, inhibitory neurons balance excitation so that runaway action potential buildup does not occur, similar to how it is theorized that the superpartner particles limit the interaction between the Higgs boson and the particles in the Standard Model so that the Higgs boson's mass is kept in check. Recall that inhibitory input to pyramidal neurons is stronger than the input from one pyramidal neuron to another because inhibitory neurons have axons with a higher bouton density; this stronger input gives inhibitory neurons a quality that is analogous to the high mass of superpartner particles. The inhibitory neurons themselves

247

might correspond to the fermion superpartners while their action potentials could correspond to the boson superpartners, similar to B-U IF's treatment of the relationship between layers 2/3 and 5 pyramidal neurons and their action potentials, i.e., Standard Model fermions that emit Standard Model bosons.

Another interesting point of note is that elements of error correction code have been found in some of the mathematical expressions of Supersymmetry.[157] These are the type of expressions that one would expect to find in computer code; some have suggested that perhaps this could be taken as evidence in support of the conjecture that the universe is actually some kind of computer simulation. Alternatively, if B-U IF is correct, this error correction code could be related to the brain's ability to perform the same function. In other words, it could be that one of the many computational talents of the inhibitory networks in the thalamocortical-hippocampal system is error correction. In this case, instead of a computer simulation, the universe would correspond to a different type of simulation, one that is carried out by our brains during spindle and sharp-wave/ripple events when newly acquired memories are being processed and integrated with pre-existing ones.

It is believed by some physicists that there may be a deep connection between the Higgs boson and dark matter, that the former may somehow provide the key to understanding the latter. This is consistent with B-U IF, because in this framework, the analogs of the Higgs boson and dark matter exist within the same structure—the thalamus—and participate in the two complementary thalamocortical circuits that coordinate and isolate neocortical areas, and provide them the fundamental level of activation to partake in neocortical information processing. Note that the lack of extensive information processing within the thalamus itself correlates well with dark matter properties, namely that compared to normal matter, it does not process information at a particularly fast pace. In

M-theoretical description of the universe will have to be used to account for the findings of special relativity. Remember that one of the key insights of relativity is that every physical system has its own internal clock; and since time is seen through change, it follows that the rate at which time elapses in a system's internal clock reflects the rate that its internal configurations change. Special relativity says that the rate of change of a system's internal and external configurations—how fast its clock ticks and how fast it moves through space, respectively— are complementary. This means that the faster a system moves through space, the slower its clock ticks and vice versa.

In what I believe to be a process analogous to string vibrations, the neuronal oscillations that are so fundamental to neocortical information processing provide a means to measure change, and hence, act as a clock to represent time. Although each neuron has its own frequency that it prefers to send and receive messages, when it is part of a coalition it becomes enslaved to the global rhythm; in this sense, the global rhythm defines a clock for the coalition. Furthermore, neurons that encode for moving objects display a property called velocity-dependent gain where their discharge rate—how fast they oscillate— increases with the speed of the object. Since these oscillations act as a clock, their variability implies that for the oscillating coalition, the passage of time is also variable and depends on the motion of the object it encodes for. All that is needed for a coalition of oscillating neurons encoding for a moving object to be completely analogous to the findings of special relativity is the assumption that as the oscillation rate increases, the amount of synchrony within the oscillating population of neurons also increases. Note that this type of increase in synchrony, or more generally, coherence, will occur at the expense of incoherent activity, which is associated with changing internal network configurations. Recall that in B-U IF, incoherent neocortical activity produces the type of feedback that corresponds to of electromagnetic radiation produced by black holes. And to completely match the description of black holes as essentially being pure gravity on the interior brane, in addition to not sending feedback to layers 2, 3, 5, and 6, the perfect synchrony of coalition members on the frontal level would also have to result in the termination of cortico-thalamocortical projections to the coalition members that once existed on the intermediate level. This is consistent with there actually not being a physical manifestation of a black hole other than the gravity that it produces; the black hole just exists on the boundary brane, or, the location of nonphysical aspects of the universe such as quantum processes and holographic information storage.

The neocortical analog to a black hole—a highly synchronous, vigorously oscillating coalition of neurons largely on the frontal level of the neocortical hierarchy, that produces strong feedback solely to layer 1 of the intermediate level capable of entraining its targeted areas—matches up well with insights emerging from M-theory and information-theoretic considerations. For instance, we saw earlier that increasingly, gravity is being looked at as the product of processes occurring within matter related to information and entropy. In the Holographic Universe framework, information stored on the boundary brane—the holographic screen of the physical universe—is the source of the entropy resulting in gravity in the interior. Therefore, for every object with mass in the interior, there are processes occurring on the boundary that result in the object's gravitational field. In this view, a black hole isn't just a singularity and an intense gravitational field; it also corresponds to a very large swarm of interacting particles—or interacting strings—on the boundary brane. Also recall that when a black hole was examined from a computational standpoint, we saw that all of its component parts function coherently as a single entity, and information is stored nonlocally within it. So we see that modern physics is moving in a direction where black holes are described as strings on the boundary brane interacting coherently to produce an intense gravitational field in the interior—a description that perfectly matches my proposed neocortical black hole analog.

So when we take the insights of B-U IF into consideration, the picture that begins to emerge is one where a system of highly correlated branes interacts to compute the universe. These branes have a structure analogous to the neocortex, namely, layered and modular, and the vibrating strings correspond to the current that courses through neocortical circuitry. One of these branes—the boundary brane—is the location of nonphysical processes, such as probability waves for quantum systems and black holes; the other brane—the interior brane—is where the physical universe manifests in the form of a holographic projection. All physical objects in the universe are computed by a coalition of strings existing on both the boundary and the interior branes. Changes to the internal state of this coalition is related to the rate of elapsed time of the internal clock of an object; furthermore, these changes result in the forces within the Standard Model, namely, the strong, weak, and electromagnetic forces. Changes in state of the charged components of these objects result in the communication of electromagnetic radiation in the form of feedback from the boundary to the interior destined for the three extended spatial dimensions. Gravity, the amount of synchrony within the coalition, is also communicated from the boundary to the interior via feedback destined for the three extended spatial dimensions.

Insights regarding the nature of quantum gravity can be obtained by considering the changes that occur to a coalition of neurons as the amount of coherent activity within it changes. In the neocortex, the coherence within a coalition is largely dependent on its size in such a way that larger coalitions tend to exhibit more coherence (a higher degree of either synchrony or constant phase and amplitude relationships between the members of the coalition). Naturally, the amount of coherent activity within a coalition of neurons is inversely proportional to the amount of incoherent activity within it so that as coherence increases, incoherence decreases. In B-U IF, this is mirrored by the shift from feedback to layers 2, 3, 5, and 6 to feedback to layer 1. B-U IF also says that a physical analog where this occurs is when an object is caught within the clutches of a black hole and is falling in. As the object approaches the black hole, it begins to travel faster and faster, approaching the speed of light. This object's internal clock begins to tick slower and slower and the light that it emits begins to get redshifted, or, stretched away. Eventually, the object falls in, at which point, its clock no longer ticks, it emits no light, and it is physically lost to the universe forever. Its mass has been added to the mass of the black hole which results in a stronger gravitational field for the black hole. Within the framework that I am proposing, we can say that as the object approached the black hole, the coalition of strings encoding for it began to oscillate more vigorously and with increasing levels of synchrony, resulting in a shift from feedback analogous to electromagnetic radiation to feedback analogous to gravity. Synchrony increased at the expense of incoherent activity (changes in internal states) which results in slower rates of elapsed time for the object's internal clock. And when it finally falls into the black hole, the only areas of the coalition still active are ones on the boundary brane; at this point, the strings encoding for the object that fell in have joined in-sync with the coalition of strings that were already encoding for the black hole. The synchronous activity of these boundary brane strings produces feedback that is destined solely for layer 1 of the interior to manifest as gravity.

In B-U IF, the coalition of strings encoding for black holes and quantum systems is confined to the boundary brane, but is still capable of sending feedback to the interior brane. In the case of black holes, this feedback is constantly applied and destined solely for layer 1, corresponding to gravity; for quantum systems, this feedback could be destined for the Higgs boson analogs in the thalamus and interior brane layers 1, 2, 3, 5, or 6 if the system is sufficiently measured. *So according to B-U IF, the amount of synchrony, or more generally, coherence, within the activity of coalitions of strings is at the heart of solving the mysteries surrounding quantum gravity.*

The Vital Spark

The dynamics of the thalamocortical-hippocampal system during NREM sleep mirrors some of the standard cosmological and M-theoretical descriptions of how universes manifest, namely, the Big Bang theory, Inflationary cosmology, and the Cyclic Universe framework. Delta frequency oscillations cause the activity of just about all excitatory and inhibitory neurons in the neocortex to become synchronized and cycle between highly active up-states and completely inactive down-states. When the state transitions from a down-state to an up-state, it can send signals to burst initiators in the thalamus that trigger thalamic spindles. The result is highly coherent spindle activity in thalamocortical circuitry that lasts for approximately 4 seconds with a frequency of about 12-14 Hz. In B-U IF, the shift from down- to up-states is the analog of the Big Bang, and the thalamocortical spindles are the analog of the Higgs ocean.

The neocortical transitions from down-states to up-states can also send signals to burst initiators in the hippocampus to trigger sharp-waves and ripples—short-lived, synchronous, and high-frequency transmissions of memory traces to the neocortex; this phenomenon happens to be one of the most synchronous network patterns observed in the brain. In B-U IF, these hippocampal signals correspond to the field that mediates cosmic inflation—the inflaton. Recall that the hippocampus is often viewed as a giant cortical column, one that stores information in a highly distributed fashion where most neurons, no matter the distance separating them, are in close contact with each other. This gives the hippocampal input—sharp-waves and ripples—the properties of the inflaton, namely, it is produced by a single unified source that is responsible for some degree of uniformity.

In B-U IF, the co-occurrence of spindles and sharp-waves/ripples corresponds to the manifestation of the universe as we know it—the brief cosmic inflation that occurs immediately after the Higgs ocean is in-place is analogous to brief hippocampal sharp-wave/ripple input at the onset of thalamocortical spindles. The co-occurrence of these spindles and sharp-waves/ripples is an example of internally-driven, self-organized activity within the brain because it occurs when there is virtually no interaction with the environment, allowing for repeated replay of neuronal sequences that occurred during waking experiences. These events peak during stage 3 sleep and create the optimum conditions within the neocortex for solidifying temporarily stored memories that have recently been acquired.

The co-occurrence of spindles and sharp-waves/ripples also resemble the Cyclic Universe framework because they happen repeatedly during NREM sleep, allowing for different memories to be processed in each successive cycle; and since the levels of neocortical hierarchy correspond to branes, their interaction with each other and with the areas of the limbic system during the transition from down- to up-states can be seen as a kind of collision, just like the collision between the branes in the Cyclic Universe framework that causes Big Bangs.

Self-similarity

After its initial moments, the universe has evolved in a way that is similar to the brain after the onset of a joint spindle and sharp-wave/ripple event. We saw earlier that the Big Bang started the universe off with a coherent introduction of matter and energy and that the quantum fluctuation believed to have sparked it provided a slight degree of variance in the initial distribution of matter. Thanks to the efforts of the four fundamental forces, this tiny variance grew into the large-scale structure of the universe, or, the cosmic web—a network of galactic structures encased by dark matter. It is important to note that these galactic structures are the locations within the universe where complexity has increased. For example, stars form here and produce the atomic elements that populate the periodic table; when a star dies, these elements are scattered about and can combine to form other celestial bodies, some of which, such as planets, can become locations where life can evolve, and as we have observed here on Earth, the appearance of life opens the door to ever increasing levels of complexity.

My principal assertion is that the human thalamocortical-hippocampal system can be used to model the universe, essentially saying that the two systems share the same structural organization and dynamics. *In this view, here on Earth, the universe has increased in complexity so much so that, with the appearance of humans, it has crossed a threshold where it displays the ultimate self-similar characteristic—it contains a subsystem that is a model of the whole.* If this is the case, it would hardly be the only instance of self-similarity displayed throughout the universe. In addition to idealized mathematical fractals, physical systems and processes with self-similar characteristics abound throughout nature and are indicative of deterministic chaos. Note that the visualization of chaotic processes often results in structures with self-similar attributes. Such chaotic processes are integral to growth within many biological systems and are responsible for the fractal patterns observed throughout all of life. It turns out that fractals are easy to create because the chaotic algorithm that produces them is simply a series of repetitive branching patterns that occur all throughout a system; furthermore, fractal patterns are very efficient components of living systems because they possess a lot of surface area, a quality that makes them good highways for communication.[25(pp79,81),26(pp8,65,76,78)]

Now let's compare these aspects of the universe to the brain at night as it processes information during NREM sleep. Like the universe, the spindle and sharp-wave/ripple events that occur during NREM also have coherent beginnings. The particular burst initiators within the thalamus and hippocampus are called up by a prominent feature of NREM sleep—the down- to up-shifts of neocortical activity occurring at delta frequencies; it is via these oscillations that neocortical neurons whose synaptic strengths have been modified by recent waking activity can activate the thalamocortical-hippocampal circuitry necessary to solidify memories. In B-U IF, the initial state of the population of neurons participating in the down- to up-shifts encodes the information that is analogous to quantum fluctuations during Big Bangs.

The evolution of the neocortical-limbic system after the initial moments of a spindle and sharp-wave/ripple event is also similar to the evolution of the universe after the Big Bang. During spindle and sharp-wave/ripple events, the sharp-waves and ripples occur very briefly at the onset of the spindles like the relationship between the inflaton and Higgs ocean. Furthermore, thalamocortical circuitry uses the entire duration of the spindle to process the hippocampal information provided briefly at spindle onset, just like all the matter that results from the inflaton allows for all of the physical processes that play out within a universal cycle. Over time, the initial high coherence existing early on within spindle and sharp-wave/ripple events will begin to decrease, giving way to increases in the complexity of neocortical dynamics—a rise in incoherence. Since only a very specific subset of thalamocortical-hippocampal circuitry is involved during these events, this complexity increase occurs in localized regions of the entire thalamocortical-hippocampal system, just as the increase in complexity is in the universe. B-U IF asserts that the complexity increase experienced by the thalamocortical-hippocampal system during spindle and sharp-wave/ripple events occurs in the same way that it does in the universe because as I've stated earlier, the structural organization of the thalamocortical system ensures that information is processed similarly to the way that it is in an M-theoretical universe, i.e., the hierarchical organization of the neocortex, and the details of its circuitry, such as lateral and feedback connections, produce processes analogous to those theorized by M-theory.

It should also be pointed out that the brain itself is one of those biological structures that feature self-similar characteristics. For one, the oscillatory dynamics of the neocortex is scale-free because small patches of it can have the same relative power distribution across the brainwave spectrum as the whole. In this sense, neocortical dynamics can lack a characteristic scale so that similar fluctuations can exist on multiple spatiotemporal scales. In addition, neuronal connectivity within the neocortex is also scale-free, and in fact, the scale-free oscillatory dynamics is in many ways a product of this scale-free connectivity. Taken together, this all means that subsets of the neocortex can produce similar patterns of activity as the neocortex as a whole, and possess similar patterns of connectivity as well.

Example: The Logistics Map and its Bifurcation Diagram

An example that may help us gain some insight as to what type of system the human brain and universe might be on a fundamental level is the logistics map, a mathematical model for the behavior of nonlinear dynamical systems.[158] The logistics map is a useful tool for describing how these types of systems evolve in time from an initial condition, and it illustrates how the short-term behavior of these systems is often well understood but their long-term behavior can become chaotic, and hence, not easily described by us due to the propagation of what were initially tiny errors in our understanding of the system's initial state.[25(p22),26(pp21,25,30)] For simplicity's sake, we will consider the discrete, one-dimensional logistics map which is defined by

$$x_{t+1} = rx_t (1-x_t).$$

The variable x represents the attractor states, or, all the states that the system is capable of manifesting. For example, in the prototypical application of this model, x stands for the population size of a particular species in an ecosystem, and the above equation models how the population size is capable of behaving as the system's nonlinearity increases, which can be brought about by changes to the environment of the species that affect its growth rate, such as changes to the population numbers of its prey. The subscripts throughout the equation represent discrete instants of time—the present time is t, and an advancement of one discrete unit of time is depicted by $t + 1$. The parameter r indicates the amount of nonlinearity displayed by the system, or, how far from equilibrium it is.[26(p42)]

The behavior of a system described by the logistics map as it becomes more and more nonlinear is illustrated by what's known as the bifurcation diagram (Figure 21). When interpreting this complex plot, it is important to keep in mind that the logistics map describes systems that are capable of manifesting more than one of its attractor states for some values of nonlinearity, r. That is, if the system has a value of r that allows for multiple attractor states, then if given enough time, the system is capable of cycling through all of these attractors states (all allowed values of x for a given r). For low r, there is just a single attractor state at any given time, but as r increases beyond a certain threshold (approximately $r = 3.0$), a bifurcation occurs so that there are now two distinct trajectories that the system could take through its state space—an indication of changes to the system's dynamics. As r continues to increase, bifurcations begin to occur more frequently along each of these paths, and eventually, the number of possible attractor states and trajectories through the system's state space becomes quite plentiful and the observed system behavior approaches deterministic chaos (approximately $r = 3.58$). All order is not lost, however, because as the system evolves through the chaotic regime, every once in a while, bands of relatively periodic activity will emerge out of the chaos (for example, approximately $r = 3.86$). On large scales of the bifurcation diagram, these bands represent a situation where few points can result in attractors. In short, one of the most important things to take away from a consideration of the bifurcation diagram is that by varying the nonlinearity of a complex dynamical system, its behavior can vary the full range from very simple to highly complex.[25(p118),26(pp47-48),159]

Another important characteristic that nonlinear dynamical systems can possess is self-similarity. For physical structures, this means that there are regions which, if magnified, will be found to be copies of the whole system. For systems and processes in general, self-similarity implies the existence of structures or processes on smaller spatiotemporal scales that possess a similar structural organization and/or dynamics as

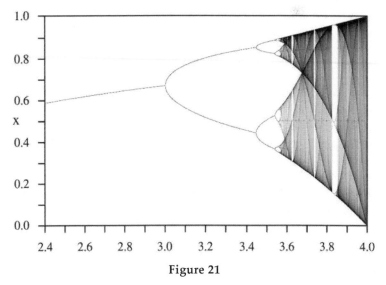

Figure 21

the whole. The bifurcation diagram contains many subregions with structure similar to the whole, qualifying it as self-similar. Furthermore, it owes its self-similarity to the ability of systems described by the logistics map—nonlinear dynamical systems—to display chaotic behavior. The bands of periodic activity within the bifurcation diagram are particularly interesting places to spot miniature copies, because when magnified, hidden bifurcation zones with similar organization as the whole can be found (Figure 22).

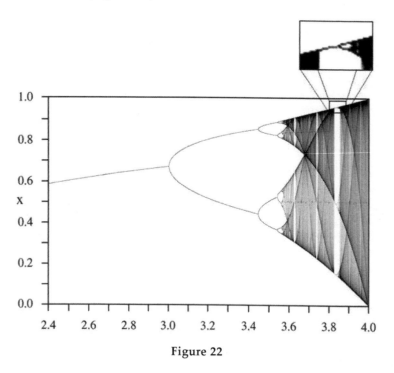

Figure 22

Complexity in the Universe Revisited

Now consider the resonance between the properties of systems described by the logistics map and the properties of the universe. The initial moments of the universe correspond to a low r condition because the high coherence of the Big Bang began things in a very simple state. Likewise, the bifurcation diagram is simplest for low values of r, when the system takes a well-defined trajectory through its state space. From the initial moments, the universe evolves in such a way that complexity increases in localized regions; analogous to the bifurcation diagram, these are locations where r increases and bifurcations begin to occur. Once far removed from the coherent beginnings, the universe's evolution can even turn chaotic (think life) so that there are regions of extremely high complexity where r has exceeded the threshold of chaos. *B-U IF asserts that in our location within the cosmic web, the universe has become so complex that a miniature copy of itself in the form of our brains has appeared.* This mirrors the appearance of mini bifurcation diagrams inside the whole bifurcation diagram after the initial onset of chaos.

There is at least one more interesting parallel between the bifurcation diagram and the universe. Recall that the Holographic Universe framework that I have been using is based on M-theory, which allows for systems of branes. These branes are arranged in a hierarchical fashion where the boundary brane is like a holographic screen encoding all of the information underlying the physical universe—the boundary brane's projection. Now notice that as r increases in the bifurcation diagram, a hierarchically-organized structure of chaotic layers begins to appear. The easiest layers to see are the ones delineated by the period three window (when the value of r is approximately 3.84 in Figure 21) but there are in reality numerous distinct layers whose apparentness depends on the scale at which the bifurcation diagram is being considered. The large chaotic zones bounding the period three window, the ones just before and just after it, could correspond to the interior and boundary branes, respectively. The sublayers within these larger zones could correspond to the extra spatial dimensions that are a part of M-theory. An interesting possibility is that the significance of the "3" in the period three window could be that it has something to do with the three extended spatial

Therefore, it is possible that within the networks participating in the spindle and sharp-wave/ripple event, the initial coherence is capable of giving way to ever increasing levels of complexity on a wide range of spatiotemporal scales, and it is possible for the complexity to increase into a chaotic regime where, at least in terms of spectral properties, what are essentially miniature copies of the whole neocortex can appear on spatiotemporal scales smaller than the whole. In this case, r has increased so much that the nonlinear dynamical system has entered a chaotic regime and has manifested its self-similar characteristics.

Now, how might the hierarchical and layered nature of the bifurcation diagram's chaotic regime apply to the hierarchical and layered nature of the neocortex? Let's segment this portion of the bifurcation diagram into three sections: 1) a low r region before the period three window; 2) a region with intermediate values of r—the period three window itself; and 3) a high r region after the period three window. The low r region could correspond to the posterior and intermediate levels of the neocortical hierarchy—the back half of the neocortex. Within this section, there are large layers that contain sublayers; in a sense, the larger layers are like the different stages of neocortical processing, such as the primary or secondary visual areas, and their sublayers are like the layers of the neocortex (layers 1-6). The same goes for the region after the period three window except that here, r is larger than it is in the region preceding the window, meaning that a higher level of complexity is capable of being manifested after the period three window. This agrees well with what we know about the neocortex, namely that the frontal areas are more complex than the posterior and intermediate areas. In the bifurcation diagram, the period three window separates the chaotic regions just before and after it, and has r values that are intermediate to these areas. The period three window holds a position that is analogous to the interface between the front and back of the brain. Comparing the bifurcation diagram and the neocortex in this way suggests that the interface between the front and back of the brain may be a critical location where order emerges out of chaos; the miniature bifurcation diagrams within the period three window appear to be symbolic of the need for the neocortex to exhibit enough complexity to manifest its self-similar nature on the neocortical level that is critical for consciousness. Furthermore, in this view, feedback from higher levels of the neocortical hierarchy is capable of modulating the r value—nonlinearity—of the targeted population of neurons on the levels below, giving the areas higher in the hierarchy an ability to induce a range of behavior that can be described as stable, periodic, or chaotic.[26(p86),158,160]

There are more similarities to be seen when we consider the flow of information in the thalamocortical-hippocampal system and the properties of the bifurcation diagram. For the lowest r region of the diagram, the trajectory through state space is approximately linear—a smooth progression through a well-defined trajectory of states. This is like the treatment of the sensory information that gets routed through the thalamus to the different regions of the neocortex; recall that the thalamus does not do much information processing on its own and is considered essentially a communications hub. One of the major recipients of thalamic information is the primary visual cortex, the lowest level of the neocortical hierarchy. From there, the flow of information splits into two streams: one stream takes the ventral route and is essential for object recognition and conscious visions, the other takes the dorsal route and mediates unconscious actions and deals with spatiotemporal information. This makes the primary visual cortex analogous to the first bifurcation in the bifurcation diagram (occurring when $r = 3$); in this analogy, the two large-scale streams of states that emerge in the bifurcation diagram as r increases are the two streams of neocortical information—ventral and dorsal. Note that in the bifurcation diagram, these streams start out distinct from each other but partially come together for intermediate values of r and completely come together for higher values of r, just like the two streams in the neocortex eventually come together in the frontal areas. What's also interesting is the hub of states that appears when the two bifurcation streams begin to partially overlap (at approx. $r = 3.68$); it is a major intersection of the two bifurcation streams and trajectories emerge from it that take the system into a regime where there is complete overlap in the two main bifurcation streams. This is similar to the central nature of the hippocampus to the flow of information in the neocortex — it is a major recipient of all sensory information from both streams of neocortical information flow, and it is well connected with the front of the brain, the most complex neocortical area and the one that receives information from both streams.

It's interesting to note that it has already been suggested that the logistics map can be used to model aspects of neocortical function such as the behavior of its fundamental modules — minicolumns and/or columns. For example, just like how the logistics map can be used to model population growth, it has also been proposed that it can be used to model network growth and dynamics in the brain.[161,162,163] No less interesting is the fact that it has also recently been shown that there is equivalence between the large-scale growth dynamics of the universe and large-scale growth dynamics of complex networks, such as the network of neurons constituting the neocortex.[17,164]

What Does It ALL Mean?

My principal hypothesis is that the human brain has the same structural organization and dynamics as the universe. After outlining qualitative models of each of these systems in Parts I and II of this book, I turned my attention here in Part III to explicitly stating some of the correspondences that exist between them. Both systems appear to be a product of the information processing being performed by a complex and dynamic network that is scale-free, self-similar, and has a hierarchical design where each level of the hierarchy is layered and consists of modular functional units. In these complex networks, there is an intricate set of connections linking the nodes, ranging from short and local projections, all the way to long distance ones connecting up widely separated parts. The systems that form are capable of manifesting anywhere from the simplest behavior, like complete synchrony, to complex behavior, like deterministic chaos. For both the brain and the universe, there are complexity levels that are really good at producing subsystems that bring out the self-similar nature of the overall system. It turns out that there is a general class of system known as nonlinear dynamical systems that have all of the qualities listed above. These systems can be modeled by the logistics map and we create the bifurcation diagram when we plot the possible attractor states that the system could have as its nonlinearity increases. This diagram provides an effective way to gain an intuitive appreciation for what types of behavior these systems are capable of displaying.

We also saw that there are already signs that the scientists who study the brain and physical universe are actually close to the realization that there are very important similarities between the respective systems that they study. On one hand, you have psychologists modeling cognition as being analogous to a quantum process, and on the other, you have physicists modeling the fabric of spacetime as if it were the same type of complex network that the neocortex is.

In the next chapter, we deal with the "so what?" of all this — why should anyone care? I make the case that B-U IF provides some of the most self-consistent and reasonable answers to questions that humanity has wrestled with throughout the ages, such as: is there a god, what's the purpose of life, and what's the true nature of reality and the theory-of-everything that describes it, if such a theory really does exist?

Chapter 15

Implications

The Power of B-U IF

If it is true that the human brain is a model of the universe, what would be the significance? What kind of impact would it have on the conclusions we arrive at when we consider our most fundamental scientific inquiries like **what is the nature of reality**? And what impact would it have on the philosophical conclusions we arrive at like **what is the nature of god, and what is the purpose of our lives**? In this chapter, I will outline some implications of B-U IF as it pertains to these questions and then compare them to what is taught in some of the more esoteric schools-of-thought, like hermetic philosophy and certain Hindu and Buddhist teachings.

The Nature of Reality

B-U IF provides a clear answer to the question: what is the nature of reality? Because it is able to do this, I feel that it provides the unifying principle needed to guide string theorists on their search for the theory-of-everything. Recall that my suggestion is that our universe is analogous to a human spindle and sharp-wave/ripple event within the thalamocortical-hippocampal system. So let's now consider what we have learned about the structural organization of the human brain and its dynamics during these spindle and sharp-wave/ripple events to see what is applicable to some of the unanswered questions plaguing string theorists on their quest for a theory-of-everything.

Calabi-Yau Shape(s)

By analyzing the equations of M-theory, string theorists have been able to limit the possible geometry of the fundamental units of spacetime to a type of shape known as a Calabi-Yau shape. The problem is that there are numerous Calabi-Yau shapes that are equally as valid and no one seems to know how to discern which shape, if any, corresponds to our universe. This limits string theory's ability to make definitive predictions; therefore, the hunt is on to identify the particular Calabi-Yau shapes that define the discrete pixels of the universe that we inhabit.

In B-U IF, the Calabi-Yau shape of our universe corresponds to the circuitry of neocortical minicolumns. Recall that neuroscientists are increasingly applying algebraic topology to the study of neocortical circuitry across a wide range of spatial scales, the smallest so far being roughly the size of two neocortical columns, and have been able to identify several multidimensional topological structures like simplices, simplicial complexes, and cavities (holes). Therefore, the answer lies in the inter- and intralaminar connections existing between the neurons of neocortical minicolumns—the fluxes within and between neocortical layers. In this framework, these fluxes allow for electrical current oscillations and action potentials to travel all throughout a minicolumn's neural network in a way that is analogous to how strings vibrate throughout a Calabi-Yau shape to produce the particles in the Standard Model. But recall that there are three generations of particles in the Standard Model, and because of this, it is expected that the Calabi-Yau shape of our universe will be found to contain three holes. Therefore, an implication of B-U IF is that there should be a corresponding set of three holes encoded within the circuitry of neocortical minicolumns. Furthermore, the way in which these three holes are encoded by neocortical neurons is determined by the depth of the neurons within neocortical layers and the patterns of their interlaminar projections. In short, B-U IF suggests that there should be three fairly-well segregated bundles of interlaminar projections within neocortical minicolumns. Note that we've already seen hints of this in the "gradient of projections" that can exist within the neocortex, such as stellate neurons at the bottom of layer 4 sending narrowly-focused projections to pyramidal neurons in layer 3, stellate neurons in the middle of layer 4 sending branching projections to pyramidal neurons situated lower within layer 3, and stellate neurons at the top of layer 4 sending profusely branching projections to pyramidal neurons situated even lower within layer 3, near the border of layers 3 and 4. In fact, this pattern of segregated interlaminar projections continues even higher up to layer 2/3, in the projections between distinct populations of layer 2/3 pyramidal neurons.[165]

Boundary Brane of our Universe

Just like string theorists do not know the particular Calabi-Yau shape that pertains to our universe, they also do not know the exact nature of the boundary brane that pertains to our universe. Recall that string theorists have developed a Holographic Universe framework where a four-dimensional, gravity-less, and point-particle quantum field theory existing on the boundary brane projects to a 10-dimensional string theory with gravity in the interior. But we know that this particular framework isn't the one that describes our universe. If string theory also applies to the boundary brane, then one might suspect that the point particles of the quantum field theory on this brane may also be represented by strings vibrating in a higher-dimensional spacetime, such as 10 dimensions; in this case, it is perhaps a 10-dimensional boundary brane that projects to a 10-dimensional interior along a dimension joining the two.

So how can we use B-U IF to gain insight about the nature of our universe's boundary brane? Consider the neocortex. It is a pretty homogeneous structure because the variety of neurons throughout it is relatively constant, and computation within local regions of it is fundamentally the same. In addition, the entire neocortex has both a layered and a modular structural organization. Of particular importance for determining the nature of our universe's boundary brane is the fact that the number of layers is the same all throughout the neocortex — six. In B-U IF, these six layers correspond to the six tiny and curled up extra dimensions within string/M-theory. Since the frontal and intermediate neocortical hierarchy levels correspond to the boundary and interior branes, respectively, neocortical homogeneity suggests that the dimensionality of the boundary brane and the interior brane are the same — 10-dimensional.

Relationship between the Branes in a Holographic Universe

In M-theory, strings vibrate within branes; furthermore, these branes can interact with each other and form systems of branes. The dimension along which they interact adds an eleventh dimension to the list of 10 previously tallied by string theory. In a Holographic Universe framework, it is along this dimension that the boundary brane projects the physical universe to the interior. With that said, the details of this projection remain unclear — exactly how is energy/matter communicated through this dimension?

To answer this question, we need to consider what the different aspects of the human neocortex correspond to in the branes of the Holographic Universe framework (Figure 23). I've already established that the frontal and intermediate levels of the neocortical hierarchy correspond to the boundary and interior branes, respectively. And I've established that the six layers and arrayed minicolumn organization of the neocortex correspond to the six tiny curled up dimensions existing at each discrete point in spacetime — the array of Calabi-Yau shapes in the branes of the Holographic Universe framework. But there is one key difference between the frontal and intermediate levels of the neocortex, suggesting that the same difference is present between the boundary and interior branes. The difference lies in the more complex structural organization of the front of the neocortex which manifests in the form of larger neurons with more extensive connectivity. But note that when it comes to complex systems, increases in complexity tend to bring about emergent properties, and one of the key emergent properties that appear to be associated with the frontal level of the neocortical hierarchy is its ability to simultaneously support numerous attractor states. Consistent with what we've seen in Part II about how the front of the brain can stealthily process much of the information underlying our conscious experiences, most of which remains hidden from our awareness, simultaneously-existing and coherent attractor states remain largely confined to the front of the brain. Only one of the simultaneously-existing attractor states at a time will be projected to the intermediate level of the neocortical hierarchy and produce conscious awareness—this phenomenon is often referred to as "the unity of consciousness". Similarly, the boundary brane in the Holographic Universe framework also appears to be able to support multiple simultaneous attractor states, a condition that is necessary for the existence of quantum phenomena, such as superposition and entanglement. It's not until one of the states

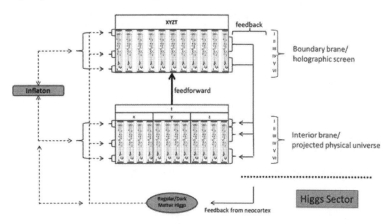

Figure 23

in the superposition is selected by sufficient measurement that a projection from the boundary brane to the interior brane manifests as a physical process.

Recall that in B-U IF, the feedback produced by the frontal level of the neocortical hierarchy destined for the intermediate level corresponds to the gravity and electromagnetic radiation that exists on the interior brane of our universe in such a way that feedback targeting layer 1 of the intermediate level is analogous to gravity while the feedback targeting layers 2, 3, 5, and 6 is analogous to electromagnetic radiation. In other words, both gravity and electromagnetic radiation arise from activity on the boundary brane and are communicated through the eleventh dimension to the interior brane as part of the projection of the physical universe. However, B-U IF suggests that there is more to the story when it comes to the ability of the boundary brane to project the physical universe to the interior. What about the fundamental amount of mass and energy associated with the matter particles in the Standard Model? In other words, how does the Higgs boson factor into the boundary brane's projection to the interior? Recall that besides direct cortico-cortical feedback, there is another way that the frontal level of the neocortical hierarchy can communicate with the intermediate level. This is through indirect connectivity with neurons in the thalamus — the network hub for the neocortex. Neurons on the frontal level of the hierarchy can send feedback projections to thalamocortical neurons whose projections target neurons on the intermediate level of the hierarchy. But also recall that some of these thalamocortical neurons correspond to the Higgs boson — the particle that confers mass to the matter particles in the Standard Model. So in this view, the boundary brane's projection not only consists of gravity and electromagnetic radiation, but also an ability to activate Higgs bosons so that matter particles can manifest in the interior. Furthermore, B-U IF suggests that the thalamus is home to not only the Higgs boson analog (specific thalamocortical neurons) but also the dark matter analog (non-specific thalamocortical neurons); therefore, B-U IF also suggests that the projection of the boundary brane can also manifest dark matter particles throughout the universe.

Quantum Gravity

Remember that in order to obtain a theory-of-everything, we will have to find one that is capable of matching the successfulness of both quantum mechanics and relativity theory at explaining the aspects of the universe that they are concerned with, and is also capable of going beyond them to explain the phenomena that they fall short on, namely, black holes and the Big Bang. Also, because the Standard Model, a quantum framework, already accounts for three out of the four fundamental forces, it is reasonable to suspect that the theory-of-everything will provide a quantum mechanical description of gravity, the one force that has yet to be accounted for by the Standard Model. We also saw that string/M-theory has proven up to this task because it can reproduce the findings of quantum mechanics and relativity theory, and it can produce candidate particles for the graviton — the theorized quantum particle that transmits the force of gravity.

In B-U IF, both electromagnetic radiation and gravity are produced by processes occurring on the boundary brane, the brane that is the home of quantum processes, like the probability waves associated with every quantum system. In this view, all physical objects are a result of the activity within a coalition of vibrating strings spread throughout both tiers of the Holographic Universe framework — the boundary and interior branes. The incoherent activity (relative oscillatory phase changes) within the coalition members on the boundary brane produces signals that are transmitted to the interior brane in the form of electromagnetic radiation. On the other hand, the synchronous activity within these boundary brane coalition members produces signals transmitted to the interior that correspond to gravity, and the graviton corresponds to the feedback action potentials that target layer 1 of the intermediate level of the neocortical hierarchy. *Therefore, I suggest that the amount of synchrony, and more generally, coherence, within a system of vibrating strings on the boundary brane will play a critical role in the quantum description of gravity.*

Information is Fundamental

If the universe corresponds to a spindle and sharp-wave/ ripple event, then it essentially is information processing; more specifically, it is a process where new information is being combined with old information. This is consistent with the recent trend among scientists to interpret the theories of physics within the context of information, viewing information as physical and using concepts from information theory to describe the communication that must take place in order for two distinct systems to interact. Since interacting parts and systems are all that exist in the universe, information theory plays a very important role in describing the nature of reality. Now physicists are starting to realize that all of their theories, such as thermodynamics, quantum mechanics, and relativity theory, are actually theories of information. If B-U IF is correct, it suggests that incorporating information theory into theoretical physics will continue to reveal much insight concerning the nature of reality and the remaining mysteries within physics.

Cyclic Occurrence of Universes

The idea that the universe corresponds to a spindle and sharp-wave/ripple event is consistent with the Cyclic Universe framework that has emerged out of M-theory. Within this framework, Big Bangs occur when two branes, interacting through the 11th dimension, collide periodically; the two branes cycle through periods of being attracted to each other, which ultimately leads to the collisions and the beginning of a new universal cycle. In this view, our universe could be one of an ensemble of universes that come into existence one after another.

Insight into these cycles can be obtained by examining the occurrence of spindle and sharp-wave/ripple events all throughout NREM sleep. Remember, these events are linked to the slow-wave oscillations that occur at delta frequency throughout NREM. When neocortical neurons switch from the silent state to the active state, they emit signals to burst initiators in the thalamus and hippocampus, thus, triggering a spindle and sharp-wave/ripple event. This suggests that the cycles of attraction in the Cyclic Model are analogous to the slow-wave modulation of neocortical neurons during NREM sleep. It also suggests that the cosmic analog to the thalamus—Higgs bosons and dark matter particles—and to the hippocampus—source of the inflaton—can be seen as constituting branes themselves. *The interaction between the neocortex, thalamus, and hippocampus during the initial moments of spindle and sharp-wave/ripple events is analogous to colliding branes in a Cyclic Universe framework.*

Other States Manifested by the Universe

Does the universe only resemble the thalamocortical-hippocampal system when a spindle and sharp-wave/ripple event is under way? Recall that human sleep consists of more than just stage 3 sleep, the stage that produces the most spindle and sharp-wave/ripple events; there are a total of five stages of sleep, including both non-conscious and semiconscious states. Perhaps somewhere within the mathematics of M-theory, there is a description of the universe that corresponds to states other than spindle and sharp-wave/ripple events. More still, the "*Cosmic Brain*" may even manifest a state analogous to the waking state, and during this time, it could function in such a way that the content of its information processing is some form of environmental stimuli and/or conscious internal cognition. For us to conclude this to be true will require us to once again expand our concept of the universe and exploit its self-similar nature. During spindle and sharp-wave/ripple events, the brain is essentially closed off from its environment, a condition that matches well with our concept of the universe—a closed system ever since the Big Bang. On the other hand, outside of purely internal, meditative-like states, the waking state implies some type of environment. So if the universe is able to manifest a state analogous to our waking state, we have to expand our concept of the universe to account for the environment of the Cosmic Brain. Furthermore, this would be another instance of self-similarity because just like we—the miniature models of the universe—exist within our environment, the universe will exist within its own environment. If the universe is truly self-similar, it would suggest that the environment which the universe senses while in its waking state is similar to our own.

Recall that five different formulations of string theory have been found and that M-theory essentially unifies them by suggesting that the universe is computed by a system of branes and that there is a parameter that can be tweaked to morph the five string theory formulations into each other. B-U IF suggests that just like the nonlinearity of regions of the neocortex can change depending on brain state, the nonlinearity of the branes in the Holographic Universe can change as well, depending on the state of the Cosmic Brain. Perhaps there are sequences of neocortical state transitions that correspond to the various string theory formulations.

Parallel Universes
If the human brain is the model of the universe, then the global human society constitutes a field of parallel universes. All of our brains exist side-by-side and they all manifest unique configurations and states, some within the waking state space,

others within the states of sleep. *For a cosmologist working within the framework of B-U IF, this paints a picture of an array of parallel universes, some in states that correspond to our current universe, others to states of the universe that we have yet to completely capture with our theories.*

One of the biggest challenges facing string theorists is determining which of the Calabi-Yau shapes out of the long list of possibilities actually correspond to our universe, because in theory, each of these shapes are equally valid. The framework that I am presenting suggests that the answer is the Calabi-Yau shapes that correspond to the geometry mapped out by human minicolumn circuitry. But does this mean that the other perfectly valid Calabi-Yau shapes have no bearing on reality? Why couldn't these shapes correspond to the minicolumns that exist in the brains of Earth's other species of animals? *There could be a whole hierarchy of universes, analogous to the hierarchy of life forms throughout the Earth (or universe perhaps) that possess a thalamocortical-hippocampal system.*

Unification

Many physicists feel that one day, when we finally obtain a theory-of-everything, we will see that all of nature's fundamental forces are actually a manifestation of one force; this singular force would become manifest only at extremely high energy levels—anything less would lead to the splintering of the force into what we experience today. In B-U IF, unification is a result of the fundamental forces being analogous to action potentials produced by neocortical neurons. Recall that once produced, these action potentials can be transmitted throughout an entire axon. These messages can exit the axon at key points such as a synapse with nearby neurons within the same, or near-by minicolumn. In addition, these messages can be sent as feedback to lower levels of the neocortical hierarchy where they can target neurons on layers 6, 5, 3, 2, or 1. In other words, once a neuron produces the message, it can travel throughout the length of the axon and influence many neurons along the way. *The locations where these interactions occur—within the same or adjacent minicolumn, or feedback destined for the various neocortical layers—is what determines if the interaction is the analog of the strong nuclear, weak nuclear, electromagnetic, or gravitational force. Furthermore, B-U IF suggests that gravity can be united with the other Standard Model forces by considering how the analogs of these forces are generated within the neocortex, namely that lateral projections and feedback to layers 6, 5, 3, and 2 are a product of incoherent activity within a coalition of neocortical neurons (change) while feedback destined for layer 1 is produced by the synchronous activity within the coalition.*

Higgs Bosons, Dark Matter, and Supersymmetry

In B-U IF, the two types of thalamocortical neurons can be viewed as analogous to two types of Higgs boson: 1) a "normal matter" Higgs boson, the one already observed at LHC; and 2) a "dark matter" Higgs that is responsible for the large-scale dark matter that shapes and molds the cosmic web. But note that both types of thalamocortical neuron are intimately connected with the network of neocortical inhibitory neurons— the analogs of the superpartners in Supersymmetry, perfectly viable candidates for another form of dark matter if found to exist. *Therefore in B-U IF, the Higgs boson is closely associated with dark matter in at least two ways because it can be considered a form of dark matter itself and it interacts directly with another form of dark matter—the superpartners.* This type of close connection between the Higgs boson and dark matter is a concept that is currently being explored within various cosmological frameworks, one of which is known as Higgsogenesis. This is yet another example of the synergy that exists between B-U IF and the concepts being explored in modern physics.

It should also be pointed out that in B-U IF, the Higgs

Brain, then it being in a non-conscious state during the manifestation of the universe most closely resembles the concept of god held by atheists—one where god does not exist. However, consider for a moment that if the correspondence between the universe and the brain extends to brain states other than spindle and sharp-wave/ripple events, such as the dreaming and waking states, then it is possible for the Cosmic Brain to make conscious decisions, some of which could in-turn affect what goes on during future manifestations of its spindle and sharp-wave/ripple events—the state that corresponds to the universe as we know it. *This represents a displacement in time between when the Cosmic Brain is capable of influencing the spindle and sharp-wave/ripples that occur within it and when those events actually occur—it can only exert its influence while in the waking state but spindle and sharp-wave/ripple events don't occur until sleep. At any rate, if the Cosmic Brain ever enters the waking state, it would be the closest thing to the most popular types of gods that have been envisioned throughout humanity, those that are conscious and play an active role in shaping what happens in the universe.* Shortly, I will consider an exception to the conscious god view such as particular Hindu and Buddhist teachings that are based on dreaming deities.

It is interesting to note that while we are in this current state of the universe, a state that is a non-conscious one for the Cosmic Brain that produces it, we ourselves—the miniature models of the universe—are in fact the closest thing to the traditional concept of god. When we are in the waking state, we can consciously make decisions that will affect the subsequent spindle and sharp-wave/ripple events that will occur when we go to sleep at night—these are the universes that occur within us. This is something that we naturally do; it is how our brains work. For instance, imagine you decide that you want to expose yourself to some novel experiences. To fulfill this desire, you decide to take a vacation in a far-off place, a region of the world that you have not yet been; let's say you choose Cambodia. Once you arrive, you will probably flip through your travel guide processing that avoids interfering with existing memories. However, the experimentally-verified instances of NREM conscious awareness discussed in Part II raise doubts about this view. So the question is: do people who achieve lucidity during NREM sleep perceive anything directly resulting from spindle and sharp-wave/ripple events? I do not have an answer to this question. However, I do believe some important distinctions can be made between a spindle and sharp-wave/ripple event during lucid NREM and one during non-lucid NREM. During lucid NREM, it is expected that there are areas of the brain that show higher levels of activation than during non-lucid NREM. One of these areas is the dorsolateral prefrontal cortex, the area on the frontal level of the neocortical hierarchy that is largely responsible for our self-reflective awareness, and the same area that is responsible for the self-awareness that manifests during a REM sleep lucid dream.[143] And just like the activity of the dorsolateral prefrontal cortex during a lucid dream can impose some level of order over neocortical dynamics, the increased activation of the dorsolateral prefrontal cortex during NREM should also constrain the possible neocortical dynamics occurring at this time. It's reasonable to expect that the more active the dorsolateral prefrontal cortex is, the less the neocortical NREM dynamics will look like they do when there is no lucidity. This applies to the spindle and sharp-wave/ ripple event—if it's even possible to achieve lucidity during this state—so that the more active the dorsolateral prefrontal cortex, the less this brain state will correspond to the universe as we know it, as I assert in B-U IF. This doesn't mean that lucidity during NREM will not yield any kind of deep insight regarding the fundamental nature of the mind or reality. In fact, I firmly believe that this was one of the critical sources of insight for the yogis and monks throughout past millennia that have made it their life's purpose to master this particular state of mind and synthesize the resultant knowledge into a potent class of worldview, one that is based on the concept of a mentally-created universe.

In short, *I assert that the amount of NREM self-awareness is inversely proportional to the amount of correspondence that exists between spindle and sharp-wave/ripple events and the current state of the universe. Therefore, if it is possible for people to become lucid during a spindle and sharp-wave/ripple event, then the more lucid they are, the less they can actually resemble a god who witnesses the evolution of the universe in all its detail.*

The Purpose of Life

B-U IF says that our universe corresponds to a spindle and sharp-wave/ripple event. When a human brain is in this state, memories are consolidated and the new ones that are stored in the hippocampus are added to existing ones in the neocortex. Our universe is an analogous process; this is a view that is consistent with the current trend to interpret physics through the lens of information. *On the most fundamental level, the universe is essentially information processing—every physical object that has ever existed or ever will, and every event that has ever happened or ever will, is a part of this information processing. Therefore, the purpose of human life is to participate in the information processing that is occurring. This is what we are doing with every activity that we engage in, whether we realize it or not.* Because humans and human society are on the high-end of the universe's complexity spectrum, humanity represents one of the more complex forms of information processing occurring throughout the universe. The most prominent example of this is the collective learning that humanity engages in, an activity that is rooted in one of the most important and fundamental functions of the human brain: to model its environment, which ultimately is at least the entire observable universe.

A key point to make here is that there are no real adjustments that need to be made by us to adhere to the purpose of life that is inferred from the universe corresponding to a spindle and sharp-wave/ripple event—just by existing, we naturally participate in the universe's information processing. We are a particular aspect of this information processing, one that resides on the high end of the universe's complexity spectrum. In this view, everything is as it should be.

Comparison with Esoteric Philosophy

After researching what is taught about the nature of reality in some of the esoteric philosophical traditions, namely hermetic philosophy and esoteric Hinduism and Buddhism, I came to the conclusion that I should meticulously investigate possible parallels between what physics says about the universe and what neuroscience says about the brain. Unlike any other reports of scientific endeavors motivated by a study of esoteric philosophy that I've seen, I decided to compare the structural organization and dynamics of the systems defined by the universe and the human brain.

In the *Kybalion*, a seminal text on hermetic philosophy written by The Three Initiates (believed to be one of the many pseudonyms used by William Walker Atkinson), it is taught that the universe is a mental creation, or, a thought produced by The ALL, an infinite cosmic mind.[166] In esoteric Hinduism and Buddhism, it is taught that the universe is the dream of a deity, such as the dream of Brahma; when this deity sleeps, it dreams and universes manifest, when it awakens, the universe as we know it no longer exists. The common theme in these teachings is that the universe is a mental creation of a deity, and a corollary is that we humans are a microcosm of the universe and/or this deity. Furthermore, *a common theme in the instructions of these esoteric traditions can be summed up as follows: if you wish to comprehend the nature of the universe, you must go "within" to comprehend the nature of the mind because the universe is a mental creation. Note that this is a subjective approach to comprehending the nature of reality; I, on the other hand, chose an objective one:*

to compare the structural organization and dynamics of the systems defined by the universe and the human brain.

My conclusion is that the universe that we perceive is analogous to at least the thalamocortical-hippocampal subsystem while it is in a spindle and sharp-wave/ripple event. *The teachings of the esoteric traditions are very similar to this, but they appear to be biased towards conscious or semiconscious states, such as conscious thoughts or dreams. Instead of these types of states, my framework suggests that the universe occurs during a non-conscious state.*

Without the type of detailed knowledge that we possess today about the structural organization and dynamics of the brain and the universe, how did the ancients develop their concept of a mentally-created and self-similar universe? I believe the answer lies in the subjective investigations that these ancient metaphysicists undertook, such as meditation and lucid dreaming.[167,168,169,170,171,172] Perhaps they excelled so much at these activities that they were able to maintain some level of consciousness within a wider array of mental states than is typically possible; if they believed that they could ascertain the nature of reality in these deeply buried mental states, they could have used their ability to maintain consciousness to take detailed notes of their experiences as the brain transitioned from awake to asleep and began cycling through the various sleep stages. *Today, it is generally believed that some brain states, such as spindle and sharp-wave/ripple events, are incompatible with conscious cognition. This could be the reason for the bias that the conclusions of ancient metaphysicists show towards conscious and semiconscious mental states — the thoughts and dreams central to their views of the nature of reality. However, even without direct access to mental phenomena during spindle and sharp-wave/ripple events, it is most likely possible that monks and yogis could still arrive at the conclusion that the universe is a mental creation based on both the patterns of mental activity that they were able to observe, and a realization of how our minds are ultimately responsible for all that we ever perceive and experience.*

The Brain-Universe Isomorphism

The isomorphism that I suggest exists between aspects of the human brain and the universe may indeed be the missing principle needed to spark the third string theory revolution and break through to a *true* theory-of-everything. The significance is that the brain provides a model system for string theorists to test their predictions out on, a luxury that they have not yet had. As one may expect, since this principle may lead to the *true* theory-of-everything, it may also provide answers to questions generally considered to be outside the realm of science and more in-line with philosophy; questions like: what is the nature of reality and god, and what is the purpose of our lives?

In B-U IF, the nature of reality is a form of information processing analogous to what is carried out by the thalamocortical-hippocampal system during a spindle and sharp-wave/ripple event, namely, a reorganization of new and old memories. Because this is most likely a non-conscious state, the Cosmic Brain should not be considered to correspond to a conscious god that actively partakes in or witnesses the events of the universe as they happen. But perhaps our concept of the universe can be expanded to include conscious waking states of this Cosmic Brain, during which time it can process information from its own environment or internal cognition. In turn, the waking experiences of this Cosmic Brain can have an effect on its future spindle and sharp-wave/ripple events— the analog of the universe as we know it. This does make it possible that this Cosmic Brain can consciously influence the types of information that is processed the next time it enters a spindle and sharp-wave/ripple event, but keep in mind that it will not have complete control because the processes involved ultimately become chaotic. *In this view, it is possible for the Cosmic*

Brain to have god-like qualities, although, these qualities are limited compared to most traditional views of god, and they manifest at a time when the universe as we know it does not.

We humans are the analogs of this Cosmic Brain and we routinely cycle through waking states and sleep states. Therefore, we are the closest thing to the concept of a conscious god—we are the gods to universes that manifest within us at night each time we go to sleep. Within the limits of our life's circumstances, we have the ability to consciously decide what types of information we want to be the subject of our future spindle and sharp-wave/ripple events. This is what we humans have been doing ever since we came into existence, and it is what we will continue to do for as long as we exist, whether we realize it or not, whether we like it or not.

Chapter 16

Conclusion

The Next Complexity Threshold

This book has been inspired by what I've been able to learn from studying the teachings of esoteric philosophy and their similarity to modern science's view of the universe. For quite some time now, this type of comparison has been a source of intrigue for many scientists and philosophers because of the amount of resonance that exists between the two systems-of-thought, particularly when quantum mechanics is compared to Eastern philosophy's metaphysics. Despite the similarities, however, the lessons offered up by esoteric traditions have yet to fully gain traction within the research being conducted in the disciplines of modern science, most notably, theoretical physics.

Esoteric philosophy is a general term that I use to describe a variety of schools-of-thought, such as hermetic philosophy, esoteric Hinduism, esoteric Buddhism, Taoism, etc. Each of these worldviews present their own unique message, but after becoming familiar with their essential lessons, I have concluded that there are a few very important common themes, themes whose joint implications I have yet to see fully explored by any of the writers and researchers who have compared esoteric philosophy to modern science thus far. These themes are that the universe is a mental creation and that we are the microcosm of the universe and/or the entity that mentally creates it. This has led me to the realization that what the esoteric traditions are really telling us is that there is an isomorphism that exists between the human brain and the universe. Therefore, I set out to investigate the existence of any similarities in the structural organization and dynamics of these two systems.

For both the universe and the brain, there are multiple

competing theories that attempt to model their structural organization and dynamics. In order for me to assess a possible correspondence between the two, I needed to first clearly define a qualitative conceptual model for each system, then I could compare the two. To build a model of the universe, I used disciplines within modern physics: quantum mechanics, relativity theory, cosmology, and string theory. To build a model of the brain, I used leading theories in neuroscience. In addition to these areas of research, I used concepts from within general systems theory, information theory, and big history.

In short, I have defined conceptual models for the universe and the brain using well established theories, as well as some of the best emerging ones. This has allowed me to find stunning correspondences between the two. The general takeaway is that the current state of the universe is analogous to a human thalamocortical-hippocampal system during a spindle and sharp-wave/ripple event. This is a condition where, indeed, the human brain can be considered to be a model of the universe, and it gives the universe a self-similar property. In fact, both the universe and the brain appear to possess self-similar and scale-free properties—the side effects of the deterministically chaotic processes that are characteristic of both systems. Within the framework that I present, the localized increase in complexity experienced throughout the universe is capable of crossing a threshold where, in forms that resemble the human brain, a model of the entire universe appears.

The isomorphism that exists between aspects of the brain and the universe has profound significance. For one, it provides a model of the universe for string theorists to test their predictions on, a luxury they have yet to experience, since by current technological standards, they cannot directly observe their predictions, and other types of sufficient physical models have yet to be found. Also, realization of the isomorphism provides possible answers to fundamental philosophical questions like:

what is the nature of reality and god, and what is the purpose of our lives?

Recall that a spindle and sharp-wave/ripple event is generally a non-conscious state. Therefore, a universe in this state should not be considered to be a conscious Cosmic Brain. In this view, there is no conscious god that plays an active role in the events of the universe as they unfold, or even a god (at least not the Cosmic Brain) that plays the role of witness or judge. However, we know that our brains can manifest states other than non-conscious ones such as the semiconscious states during dream sleep and fully conscious ones during waking life. If the correspondence between the universe and the brain extends into these types of states, then perhaps the Cosmic Brain can become conscious and exert some control over the types of experiences it has within its own environment or cognitive space, experiences that in-turn will determine what information gets processed during subsequent spindle and sharp-wave/ripple events. This, of course, would not be an ability of the Cosmic Brain to exert total control over what occurs in its spindle and sharp-wave/ripple events because presumably this Cosmic Brain would, like us, be constrained by the circumstances of its environment. Furthermore, the evolution of spindle and sharp-wave/ripple events turns chaotic in the long run, further limiting any ability to determine events beforehand. At any rate, if the Cosmic Brain can manifest a conscious waking state, then it can possess qualities that make it similar to a conscious god—it just so happens that these qualities do not manifest at the same time that the universe we are familiar with does.

On the other hand, we ourselves are the closest concept to a conscious god since, while in the waking state, we can consciously select the types of information that get processed during our subsequent spindle and sharp-wave/ripple events, the brain state that is analogous to universes—the universes within us. *But note that the sum total of our experiences is what* *defines who we are as individuals and plays a large role in determining the trajectory of our lives. Therefore, like two sides of the same coin, increased conscious selection of the types of information that gets processed during our future spindle and sharp-wave/ripple events is equivalent to increased conscious influence over the types of individuals we become, or, the trajectory that our lives take as they continue to play out and evolve. Increased conscious influence over the universes that manifest within us during spindle and sharp-wave/ripple events occurs naturally during an activity called character building—the **aligning** of our interests, attentional focus, knowledge acquisition, ethics, and actions, as well as the **active cultivation** of our desires, to become the types of individuals that we want to be.*[173]

A human spindle and sharp-wave/ripple event is a state where temporarily-stored memories in the hippocampus are transferred to the neocortex to be incorporated into the network of long-term memories; this is a form of information processing. Therefore, in B-U IF, the processes of the universe are also a form of information processing. The lives we live are our contribution to this universal process. Humans represent a complex form of this information processing. We mark a critical complexity threshold crossed by the universe within the cosmic web of matter where there is now a physical model of the universe existing within it. The human brain not only models the universe physically, but its function is to create mental models of it as well. Therefore, we are central to the continued complexification of the universe because of our capacity to learn collectively, and because of the society that we create as a consequence.

When life appeared on this planet, the process of biological evolution commenced. Throughout this process, life has tried, and continues to try, to improve upon how well it adapts to its environment, oftentimes leading to ever more complex forms. For instance, conditions on Earth have remained favorable long enough for life to produce humans—as far as we know, the most complex life form in the known universe. And with our appearance, a new type of evolution has commenced, one that occurs not biologically, but conceptually, involving the systems-of-thought that we collectively develop over time. Together, these systems-of-thought are like our models for everything that we've ever thought of, such as how to survive off the land, how to avoid predators, how to interact with different tribes, how to create beautiful abstract forms, how the mind works, etc. More relevant to the current discussion are the systems-of-thought that are concerned with the nature of reality, namely religion, philosophy, and science. These particular systems-of-thought are evolving along a trajectory that is leading us to the realization that our brain, the organ central to our ability to experience the world, is the model of the universe. This is the fundamental principle that will eventually pave the way for the development of the theory-of-everything. *Once this theory has fully formed, it will hold a spot within our hierarchy of systems-of-thought that is analogous to the spot held by humans in the animal kingdom.* Acknowledgement of the isomorphism between aspects of our brains and the universe will be akin to a state of lucidity—a situation where the model now realizes that it is the model. And like all of the useful concepts that we have stumbled upon in the past, we may one day learn how to use the knowledge that our brains can be used to model the universe to further transform society and our relationship to our environment. In short, this realization would mark the crossing of another critical complexity threshold by the ever-evolving universe. A threshold that will allow you to *"be-you if..."* you so-choose to cross it.

References

1. Capra, F. *The Tao of Physics: An Exploration of the Parallels between Modern Physics and Eastern Mysticism*, 25th Anniversary Edition. Boston: Shambhala, 2000.

2. Talbot, M. *The Holographic Universe*. New York: Harper Perennial, 1992.

3. Lanza, R. and Berman, B. *Biocentrism: How Life and Consciousness are the Keys to Understanding the True Nature of the Universe (audible version)*. Audible Studios, 2009.

4. Kastrup, B. *The Idea of the World: A multi-disciplinary argument for the mental nature of reality*. Alresford: iff Books, 2019.

5. Koch, C. *The Quest for Consciousness: A Neurobiological Approach*. Englewood, CO: W.H. Freeman, 2004.

6. Conant, R.C., & Ashby, W.R. "Every Good Regulator of a System Must be a Model of that System." *Int. J. Systems Sci.* **1**(2): 89-97 (1970).

7. Frith, C. *Making up the Mind: How the Brain Creates Our Mental World*. Malden, MA: Blackwell Publishing, 2007.

8. Meyniel, F. and Dehaene, S. "Brain networks for confidence weighting and hierarchical inference during probabilistic learning." *Proc Natl Acad Sci USA* **114**(19): E3859-E3868 (2017).

9. Brunton, B.W. et al. "Rats and humans can optimally accumulate evidence for decision-making." *Science* **340**(6128): 95-8 (2013).

10. Ernst, M.O. and Banks, M.S. "Humans integrate visual and haptic information in a statistically optimal fashion." *Nature* **415**(6870): 429-33 (2002).

11. Yao, T. et al. "Saccade-synchronized rapid attention shifts in macaque visual cortical area MT." *Nature Communications* **9**(958): (2018).

12. Banino, A. et al. "Vector-based navigation using grid-like representations in artificial agents." *Nature* **557**(7705): 429-433 (2018).

13. Gulyas, A. et al. "Navigable networks as Nash equilibria of navigation games." *Nature Communications* **6**(7651): (2015).

14. Geisler, W.S. "Ideal Observer Analysis." In *The Visual Neurosciences*, edited by L.M. Chalupa and J.S. Werner, pp825-837. MIT Press, 2003.

15. Moskowitz, C. "Higgs Boson Predictors Awarded the 2013 Nobel Physics Prize." *Scientific American* (2013, October 8). https://www.scientificamerican.com/article/nobel-physics-prize-higgs-englert/

16. Duff, M. "Theory of Everything." *New Scientist* **2815**: (2011).

17. Krioukov, D. et al. "Network Cosmology." *Nature — Scientific Reports* **2**(793): (2012).

18. Kvam, P.D. et al. "Interference effects of choice on confidence: Quantum characteristics of evidence accumulation." *PNAS* **112**(34): 10645-10650 (2015).

19. Busemeyer, J.R. and Wang, Z. "What Is Quantum Cognition, and How Is It Applied to Psychology?" *Current Directions in Psychological Science* **24**(3): 163-169 (2015).

20. Bruza, P.D. et al. "Quantum cognition: a new theoretical approach to psychology." *Trends in Cognitive Sciences* **19**(7): 383-393 (2015).

21. Greene, B. *The Hidden Reality: Parallel Universes and the Deep Laws of the Cosmos*. New York: Alfred A. Knopf, 2011.

22. Brooks, M. "The limits of knowledge: Things we'll never understand." *New Scientist* **2811**: (2011).

23. "The Elegant Universe." *NOVA*. Performed by B. Greene. 2004.

24. Popkin, G. 2016. "Physicists, the Brain is Calling You." *APS News*, April: 1-3. https://www.aps.org/publications/apsnews/201604/brain.cfm

25. Çambel, A.B. *Applied Chaos Theory: A Paradigm for*

Complexity. San Diego: Academic Press, 1993.

26. Strogatz, S. *Chaos* (The Great Courses series—Course and Guidebook). Chantilly, VA: The Great Courses, 2008.

27. Reimann, M.W. et al. "Cliques of Neurons Bound into Cavities Provide a Missing Link between Structure and Function." *Frontiers in Computational Neuroscience* **11**(48): (2017).

28. Sizemore, A.E., C. Giusti, A. Kahn, J.M. Vettel, R.F. Betzel, and D.S. Bassett. "Cliques and Cavities in the Human Connectome." *J. Comp. Neurosci.* **44**(1): 115-145 (2018).

29. Sizemore, A.E., J.E. Phillips-Cremins, R. Ghrist, and D.S. Bassett. "The importance of the whole: Topological data analysis for the network neuroscientist." *Network Neuroscience* **3**(3): 656-673 (2019).

30. Wolchover, N. "A Jewel at the Heart of Quantum Physics." *Quanta Magazine* (September 2013). https://www.quantamagazine.org/physicists-discover-geometry-underlying-particle-physics-20130917/

31. Ananthaswamy, A. "The new shape of reality." *New Scientist* **3136**: (2017).

32. Montiel, M.E., A.S. Aguado, and E. Zaluska. "Topology in Fractals." *Chaos, Solitons & Fractals* **7**(8): 1187-1207 (1996).

33. Song, C., S. Havlin, and H.A. Makse. "Self-similarity of complex networks." *Nature* **433**(7024): 392-395 (2005).

34. Song, C., S. Havlin, and H.A. Makse. "Origins of fractality in the growth of complex networks." *Nature Physics* **2**: 275-281 (2006).

35. Chaisson, E.J. "Energy Rate Density as a Complexity Metric and Evolutionary Driver." *Complexity* **16**(3): 27-40 (2010).

36. Bennett, K. "The Chaos Theory of Evolution." *New Scientist* **2782**: 2010. https://www.newscientist.com/article/mg20827821-000-the-chaos-theory-of-evolution/

37. Skyttner, L. *General Systems Theory: Problems, Perspectives, Practice*. Hackensack, NJ: World Scientific Publishing Co.,

2005.

38. Gefter, A. "Is the universe a fractal?" *New Scientist* **2594**: 30-33 (2007).

39. Greene, B. *The Fabric of the Cosmos: Space, Time, and the Texture of Reality*. New York: Alfred A. Knopf, 2004.

40. Close, F. *The New Cosmic Onion: Quarks and the Nature of the Universe*. Boca Raton, FL: CRC Press, 2007.

41. Helmholtz Association of German Research Centres. "Twelve matter particles suffice in nature." Phys.org, 2012 (http://phys.org/news/2012-12-particles-suffice-nature. html).

42. Riordan, M. and Zajc, W.A. "The First Few Microseconds." *Scientific American* **294**(5): 34-41 (2006).

43. Close, F. "Of arrows and eternity." *Physics World* **23**(6): (2010).

44. Shiga, D. "Beware Higgs impostors at the LHC." *New Scientist* **2812**: (2011).

45. Feng, J. and Trodden, M. "Dark Worlds." *Scientific American* **303**(5): 38-45 (2010).

46. Slezak, M. "Higgs boson having an identity crisis." *New Scientist* **2896**: 2010 (https://www.newscientist.com/article/ dn23003-higgs-boson-having-an-identity-crisis/).

47. Bekenstein, J.D. "Information in the Holographic Universe." *Scientific American* **289**(2): 58-65 (2003).

48. Seife, C. *Decoding the Universe: How the New Science of Information Is Explaining Everything in the Cosmos, from Our Brains to Black Holes*. New York: Penguin Books, 2006.

49. Christian, D. *Big History: The Big Bang, Life on Earth, and the Rise of Humanity* (The Great Courses series—Course and Guidebook). Chantilly, VA: The Great Courses, 2008.

50. Jamieson, V. "Reality: The bedrock of it all." *New Scientist* **2884**: (2012).

51. Ananthaswamy, A. "Quantum shadows: The mystery of matter deepens." *New Scientist* **2898**: (2013).

52. Mullins, J. "When the multiverse and many-worlds collide." *New Scientist* **2815**: (2011).

53. Webb, R. "Is quantum theory weird enough for the real world?" *New Scientist* **2774**: (2010).

54. Bransden, B.H. and Joachain, C.J. *Quantum Mechanics 2nd edition*. Prentice Hall, 2000.

55. Whalen, A.D. *Detection of Signals in Noise*. New York: Academic Press, 1971.

56. Downing, J.J. *Modulation Systems and Noise*. United Kingdom: Prentice Hall International, 1964.

57. Eldar, Y.C. and Oppenheim, A.V. "Quantum Signal Processing." *IEEE Signal Processing Magazine* **19**(6): 12-32 (November 2002).

58. Thron, C. and Watts, J. "A Signal Processing Interpretation of Quantum Mechanics." (https://arxiv.org/abs/1205.1681), 2012.

59. Siegried, T. "A new view of gravity: Entropy and information may be crucial concepts for explaining roots of familiar force." *Science News* **178**(7): 26-29 (2010).

60. Jacobson, T.A. and Parentani, R. "An Echo of Black Holes." *Scientific American* **293**(6): 68-75 (2005).

61. Lloyd, S. "The digital universe." *Physics World* **21**(11): 30-36 (2008).

62. Chown, M. "Our World May be a Giant Hologram." *New Scientist* **2691**: 24-27 (2009).

63. Shiga, D. "The cosmos—*before* the big bang." *New Scientist* **2601**: 28-33 (2007).

64. Grossman, L. "Turbulent black holes grow fractal skins as they feed." *New Scientist* **2967**: (2014).

65. Loeb, A. and Pritchard, J. "The universe: the full story." *New Scientist* **2888**: (2012).

66. SubbaRao, M. and Aragon-Calvo, M. "A picture of the cosmos." *Physics World* **21**(12): 29-32 (2008).

67. Coutinho, B.C. et al. "The Network Behind the Cosmic

Web." (http://arxiv.org/abs/1604.03236), 2016.

68. Scrimgeour, M.I. et al. "The WiggleZ Dark Energy Survey: the transition to large-scale cosmic homogeneity." *Monthly Notices of the Royal Astronomical Society Journal* **425**(1): 116-134 (2012).

69. Clark, S. "Dark energy: seeking the heart of darkness." *New Scientist* **2591**: (2007).

70. Aron, J. "Largest structure challenges Einstein's smooth cosmos." *New Scientist* **2900**: (2013).

71. The Kavli Foundation. "Are we closing in on dark matter?" Phys.org, 2012 (http://phys.org/news/2012-12-dark.html).

72. Chalmers University of Technology. "Higgs particle can disintegrate into particles of dark matter, according to new model." Phys.org, 2015 (http://phys.org/news/2015-03-higgs-particle-disintegrate-particles-dark.html).

73. Calder, L. and Lahav, O. "Dark energy: how the paradigm shifted." *Physics World* **23**(1): 32-38 (2010), http://physicsworld.com/cws/article/print/2010/jun/02/dark-energy-how-the-paradigm-shifted.

74. Krauss, L.M. and Scherrer, R.J. "The End of Cosmology?" *Scientific American* **298**(3): 46-53 (2008).

75. Davis, T.M. "Is the Universe Leaking Energy?" *Scientific American* **303**(1): 38-47 (2010).

76. Coles, P. "Boomtime." *New Scientist* **2593**: 33-37 (2007).

77. Maldacena, J. "The Illusion of Gravity." *Scientific American* **293**(5): 56-63 (2005).

78. Merali, Z. "The Universe is a string-net liquid." *New Scientist* **2595**: (2007).

79. Stanford Continuing Studies Program (Susskind, L.). "String Theory and M-Theory" (lecture 3). iTunes U. 2011.

80. Bars, I. and Rychkov, D. "Is String Interaction the Origin of Quantum Mechanics?" *Physics Letters B* **739**: 451-456 (2014).

81. Gefter, A. "What kind of bang was the big bang?" *New Scientist* **2871**: (2012).

82. Smolin, L. "The unique universe." *Physics World* **22**(6): 21-26 (June 2009).

83. Krauss, L. "Why one Higgs boson will not be enough." *The Guardian*, 2011 (https://www.theguardian.com/science/2011/dec/13/one-higgs-boson-not-enough).

84. CERN. http://home.web.cern.ch/about/physics/supersymmetry.

85. Servant, G., Tulin, S. "Baryogenesis and Dark Matter through a Higgs Asymmetry." *Phys. Rev. Lett.* **111**(15): 2013.

86. Zyga, L. "Could 'Higgsogenesis' explain dark matter?" Phys. org, 2013 (http://phys.org/news/2013-10-higgsogenesis-dark.html).

87. Gleiser, M. "The imperfect universe: Goodbye, theory of everything." *New Scientist* **2759**: (2010).

88. Bostrom, N. "Do we live in a computer simulation?" *New Scientist* **2579**: 38-39 (2006).

89. Brooks, M. "Reality: It's nothing but information." *New Scientist* **2884**: (2012).

90. Battersby, S. "The theory of everything: Are we nearly there yet?" *New Scientist* **2497**: 30-34 (2005).

91. Ulmer, S. "Big Bang under the microscope." Phys.org, 2013 (http://phys.org/news/2013-01-big-microscope.html).

92. Salk, J. *Anatomy of Reality: Merging of Intuition and Reality (Convergence)*. New York: Praeger Publishers Inc., 1985.

93. Spinney, L. "Busted! The myth of technological progress." *New Scientist* **2884**: (2012).

94. Clauset, A. et al. "Hierarchical structure and the prediction of missing links in networks." *Nature* **453**(7191): 98-101 (2008).

95. Santa Fe Institute. "Was life inevitable? New paper pieces together metabolism's beginnings." Phys.org, 2012 (http://phys.org/news/2012-12-life-inevitable-paper-pieces-metabolism.html).

96. Lane, N. "Life: is it inevitable or just a fluke?" *New Scientist* **2870**: (2012).

97. Vedral, V. "The surprise theory of everything." *New Scientist* **2886**: (2012).

98. Marshall, M. "Climate change determined humanity's global conquest." *New Scientist* **2883**: (2012).

99. Bettencourt, L. "The hidden laws that govern city living." *New Scientist* **2947**: 30-31 (2013).

100. Penrose, R. "The Big Questions: What is reality?" *New Scientist* **2578**: (2006).

101. Buchanan, M. "Social networks: The great tipping point test." *New Scientist* **2770**: (2010).

102. Robson, D. "The mind's eye: How the brain sorts out what you see." *New Scientist* **2775**: (2010).

103. Buzsáki, G. *Rhythms of the Brain*. New York: Oxford University Press, 2006.

104. Marshall, J. "Future recall: your mind can slip through time." *New Scientist* **2596**: (2007).

105. Collins, C.E. and Kaas, J.H. *The Primate Visual System (Frontiers in Neuroscience)*. Boca Raton, FL: CRC Press, 2004.

106. Mallot, H.A. *Computational Vision: Information Processing in Perception and Visual Behavior 2nd edition*. MIT Press, 2000.

107. Peters, A. and Rockland, K.S. *Cerebral Cortex: Volume 10, Primary Visual Cortex in Primates*. New York: Plenum Press, 1994.

108. LaBerge, D. and Kasevich, R. "The apical dendrite theory of consciousness." *Neural Networks* **20**(9): 1004-1020 (2007).

109. Grinvald, A. and Hildesheim, R. "VSDI: A New Era in Functional Imaging of Cortical Dynamics." *Nature Reviews, Neuroscience* **5**(11): 874-885 (2004).

110. Thomson, H. "Alpha, beta, gamma: The language of brainwaves." *New Scientist* **2768**: (2010).

111. Traub, R.D. et al. "A mechanism for generation of long-range synchronous fast oscillations in the cortex." *Nature*

383(6601): 621-624 (1996).

112. Katz, P.S. "Synaptic gating: the potential to open closed doors." *Curr Biol.* **13**(14): R554-6 (2003).

113. Bukalo, O. et al. "Synaptic plasticity by antidromic firing during hippocampal network oscillations." *Proc Natl Acad Sci USA* **110**(13): 5175-80 (2013).

114. Berg, R.W. et al. "Comment on 'Penetration of Action Potentials During Collision in the Median and Lateral Giant Axons of Invertebrates'." *Physical Review X* **7**: (2017).

115. Wilson, N.R. et al. "Division and subtraction by distinct cortical inhibitory networks *in vivo.*" *Nature* **488**(7411): 343-348 (2012).

116. Ananthaswamy, A. "Throwing shapes." *New Scientist* **3145**: (2017).

117. Thomson, A.M. "Neocortical layer 6, a review." *Frontiers in Neuroanatomy* **4**(13): (2010).

118. Mountcastle, V.B. "The columnar organization of the neocortex." *Brain* **120**(4): 701-722 (1997).

119. VanRullen, R. and Koch, C. "Is perception discrete or continuous?" *Trends in Cognitive Sciences* **7**(5): 207-213 (2003).

120. Joliot, M. et al. "Human oscillatory brain activity near 40 Hz coexists with cognitive temporal binding." *PNAS* **91**(24): 11748-11751 (1994).

121. Issa, N.P. "Models and Measurements of Functional Maps in V1." *Journal of Neurophysiology* **99**(6): 2745-2754 (2008).

122. Wu, J.Y. et al. "Propagating Waves of Activity in the Neocortex: What They Are, What They DO." *The Neuroscientist* **14**(5): 487-502 (2008).

123. Ermentrout, G.B. and Kleinfeld, D. "Traveling Electrical Waves in Cortex: Insights from Phase Dynamics and Speculation on a Computational Role." *Neuron* **29**(1): 33-44 (2001).

124. Llinás, R. and Ribary, U. "Coherent 40-Hz oscillation

characterizes dream state in humans." *PNAS* **90**(5): 2078-2081 (1993).

125. Ahmed, O.J. et al. "Reactivation in Ventral Striatum during Hippocampal Ripples: Evidence for the Binding of Reward and Spatial Memories?" *Journal of Neuroscience* **28**(40): 9895-9897 (2008).

126. Dvorak-Carbone, H. and Schuman, E.M. "Long-term depression of temporoammonic-CA1 hippocampal synaptic transmission." *Journal of Neurophysiology* **81**(3): 1036-1044 (1999).

127. Cashdollar, N. et al. "Hippocampus-dependent and -independent theta-networks of active maintenance." *PNAS* **106**(48): 20493-20498 (2009).

128. Marshall, L. and Born, J. "The contribution of sleep to hippocampus-dependent memory consolidation." *Trends in Cognitive Sciences* **11**(10): 442-450 (2007).

129. Jenson, O. "Reading the hippocampal code by theta phase-locking." *Trends in Cognitive Sciences* **9**(12): 551-553 (2005).

130. Molle, M. and Born, J. "Hippocampus Whispering in Deep Sleep to Prefrontal Cortex—For Good Memories?" *Neuron* **61**(4): 496-498 (2009).

131. Sirota, A. et al. "Communication between neocortex and hippocampus during sleep in rodents." *PNAS* **100**(4): 2065-2069 (2003).

132. Doeller, C.F. et al. "Evidence for grid cells in human memory network." *Nature* **463**(7281): 657-661 (2010).

133. Owen, A. and Highfield, R. "Putting your intelligence to the ultimate test." *New Scientist* **2784**: 38-43 (2010).

134. Colby, C.L. and Goldberg, M.E. "Space and Attention in Parietal Cortex." *Annu. Rev. Neurosci.* **22**: 319-349 (1999).

135. Haxby, J.V. et al. "Distinguishing the functional roles of multiple regions in distributed neural systems for visual working memory." *Neuroimage* **11**(5, pt1): 380-391 (2000).

136. Hamzelou, J. "Want to find your mind? Learn to direct

your dreams." *New Scientist* **2764**: (2010).

137. Stickgold, R. et al. "Sleep, Learning, and Dreams: Off-line Memory Reprocessing." *Science* **294**(5544): 1052-1057 (2001).

138. Liebe, S. et al. "Theta coupling between V4 and prefrontal cortex predicts visual short-term memory performance." *Nature Neuroscience* **15**(3): 456-462 (2012).

139. Raichle, M.E. "The Brain's Dark Energy." *Scientific American* **302**(3): 44-49 (2010).

140. Fisher, R. "Daydream your way to creativity." *New Scientist* **2869**: (2012).

141. Ananthaswamy, A. "Firing on all neurons: Where consciousness comes from." *New Scientist* **2752**: (2010).

142. Itti, L. and Koch, C. "Computational Modelling of Visual Attention." *Nature* **2**(3): 194-203 (2001).

143. Stumbrys, T. and Erlacher, D. "Lucid dreaming during NREM sleep: Two case reports." *International Journal of Dream Research* **5**(2): 151-155 (2012).

144. Siclari, F. et al. "The neuronal correlates of dreaming." *Nature Neuroscience* **20**(6): (2017).

145. Mason, L.I. et al. "Electrophysiological Correlates of Higher States of Consciousness During Sleep in Long-Term Practitioners of the Transcendental Meditation Program." *Sleep* **20**(2): 102-110 (1997).

146. Siapas, A.G. and Wilson, M.A. "Coordinated Interactions between Hippocampal Ripples and Cortical Spindles during Slow-Wave Sleep." *Neuron* **21**(5): 1123-1128 (1998).

147. Zygierewicz, J. et al. "High resolution study of sleep spindles." *Clinical Neurophysiology* **110**(12): 2136-2147 (1999).

148. Werth, E. et al. "Spindle frequency activity in the sleep EEG: individual differences and topographic distribution." *Electroencephalography and Clinical Neurophysiology* **103**(5): 535-542 (1997).

149. Eschenko, O. et al. "Elevated sleep spindle density after learning or after retrieval in rats." *Journal of Neuroscience* **26**(50): 12914-12920 (2006).

150. Ji, D. and Wilson, M.A. "Coordinated memory replay in the visual cortex and hippocampus during sleep." *Nature Neuroscience* **10**(1): 100-107 (2007).

151. Clemens, Z. et al. "Temporal coupling of parahippocampal ripples, sleep spindles, and slow oscillations in humans." *Brain* **130**(11): 2868-2878 (2007).

152. Stickgold, R. et al. "Replaying the Game: Hypnagogic Images in Normals and Amnesics." *Science* **290**(5490): 350-353 (2000).

153. DeFelipe, J. "The anatomical problem posed by brain complexity and size: a potential solution." *Front. Neuroanat.* **9**(104): (2015).

154. Szabadics, J. et al. "Excitatory effect of GABAergic axo-axonic cells in cortical microcircuits." *Science* **311**(5758): 233-235 (2006).

155. Khirug, S. et al. "GABAergic depolarization of the axon initial segment in cortical principal neurons is caused by the Na-K-2Cl cotransporter NKCC1." *J Neurosci* **28**(18): 4635-4639 (2008).

156. Woodruff, A. et al. "Depolarizing Effect of Neocortical Chandelier Neurons." *Frontiers in Neural Circuits* **3**(15): (2009).

157. Gates, J. "Symbols of Power." *Physics World* **23**(6): 34-39 (2010).

158. Nandi, A. et al. "The phase-modulated logistics map." *Chaos* **15**(2): (2005).

159. Hellen, E.H. "Real-time finite difference bifurcation diagrams from analog electronic circuits." *Am. J. Phys.* **72**(4): 499-502 (2004).

160. Pyragas, K. "Control of chaos via extended delay feedback." *Physics Letters A* **206**(5-6): 323-330 (1995).

161. Farhat, N.H. "Corticonic models of brain mechanisms underlying cognition and intelligence." *Physics of Life Reviews* **4**(4): 223-252 (2007).

162. Farhat, N.H. "Biomorphic Dynamical Networks for Cognition and Control." *Journal of Intelligent and Robotic Systems* **21**(2): 167-177 (1998).

163. Pashaie, R. and Farhat, N.H. "Self-Organization in a Parametrically Coupled Logistic Map Network: A Model for Information Processing in the Visual Cortex." *IEEE Transactions on Neural Networks* **20**(4): 597-608 (2009).

164. Bianconi, G. and Rahmede, C. "Complex Quantum Network Manifolds in Dimension $d>2$ are Scale-Free." *Nature Scientific Reports* **5**: (2015).

165. Yoshimura, Y. et al. "Excitatory cortical neurons form fine-scale functional networks." *Nature* **433**(7028): 868-73 (2005).

166. The Three Initiates. *The Kybalion*. Chicago: Yogi Publication Society, 1912.

167. Rock, A. *The Mind at Night: The New Science of How and Why We Dream*. New York: Basic Books, 2005.

168. Evans-Wentz, W.Y. *Tibetan Yoga and Secret Doctrines: Seven Books of Wisdom of the Great Path*. New York: Oxford University Press, 2000.

169. David-Neel, A. *Magic and Mystery in Tibet*. New York: Dover Publications, 1971.

170. Govinda, L.A. *Foundations of Tibetan Mysticism*. York Beach, ME: Red Wheel/Weiser Books, 1969.

171. Harary, K. *Lucid Dreams in 30 Days: The Creative Sleep Program*. New York: St. Martin's Griffin, 1999.

172. Satchidananda, S.S. *The Yoga Sutras of Patanjali*. Boston: Shambhala Publications, 2012.

173. Dumont, T.Q. *The Master Mind: Or the Key to Mental Power, Development and Efficiency*. Andesite Press, 2015.

IFF
BOOKS

ACADEMIC AND SPECIALIST

Iff Books publishes non-fiction. It aims to work with authors and titles that augment our understanding of the human condition, society and civilisation, and the world or universe in which we live.
If you have enjoyed this book, why not tell other readers by posting a review on your preferred book site.

Recent bestsellers from Iff Books are:

Why Materialism Is Baloney
How true skeptics know there is no death and fathom answers to life, the universe, and everything
Bernardo Kastrup
A hard-nosed, logical, and skeptic non-materialist metaphysics, according to which the body is in mind, not mind in the body.
Paperback: 978-1-78279-362-5 ebook: 978-1-78279-361-8

The Fall
Steve Taylor
The Fall discusses human achievement versus the issues of war, patriarchy and social inequality.
Paperback: 978-1-78535-804-3 ebook: 978-1-78535-805-0

Brief Peeks Beyond
Critical essays on metaphysics, neuroscience, free will,
skepticism and culture
Bernardo Kastrup
An incisive, original, compelling alternative to current mainstream
cultural views and assumptions.
Paperback: 978-1-78535-018-4 ebook: 978-1-78535-019-1

Framespotting
Changing how you look at things changes how
you see them
Laurence & Alison Matthews
A punchy, upbeat guide to framespotting. Spot deceptions and
hidden assumptions; swap growth for growing up. See and be free.
Paperback: 978-1-78279-689-3 ebook: 978-1-78279-822-4

Is There an Afterlife?
David Fontana
Is there an Afterlife? If so what is it like? How do Western ideas
of the afterlife compare with Eastern? David Fontana presents the
historical and contemporary evidence for survival of
physical death.
Paperback: 978-1-90381-690-5

Nothing Matters
a book about nothing
Ronald Green
Thinking about Nothing opens the world to everything by
illuminating new angles to old problems and stimulating new
ways of thinking.
Paperback: 978-1-84694-707-0 ebook: 978-1-78099-016-3

Panpsychism
The Philosophy of the Sensuous Cosmos
Peter Ells
Are free will and mind chimeras? This book, anti-materialistic but respecting science, answers: No! Mind is foundational to all existence.
Paperback: 978-1-84694-505-2 ebook: 978-1-78099-018-7

Punk Science
Inside the Mind of God
Manjir Samanta-Laughton
Many have experienced unexplainable phenomena; God, psychic abilities, extraordinary healing and angelic encounters. Can cutting-edge science actually explain phenomena previously thought of as 'paranormal'?
Paperback: 978-1-90504-793-2

The Vagabond Spirit of Poetry
Edward Clarke
Spend time with the wisest poets of the modern age and of the past, and let Edward Clarke remind you of the importance of poetry in our industrialized world.
Paperback: 978-1-78279-370-0 ebook: 978-1-78279-369-4

Readers of ebooks can buy or view any of these bestsellers by clicking on the live link in the title. Most titles are published in paperback and as an ebook. Paperbacks are available in traditional bookshops. Both print and ebook formats are available online. Find more titles and sign up to our readers' newsletter at http://www.johnhuntpublishing.com/non-fiction
Follow us on Facebook at
https://www.facebook.com/JHPNonFiction
and Twitter at https://twitter.com/JHPNonFiction